T0210551

Lecture Notes
in Business Information Processing 250

More information about this series at http://www.springer.com/series/7911

Shaofeng Liu · Boris Delibašić
Festus Oderanti (Eds.)

Decision Support Systems VI – Addressing Sustainability and Societal Challenges

2nd International Conference, ICDSST 2016
Plymouth, UK, May 23–25, 2016
Proceedings

 Springer

Editors
Shaofeng Liu
Graduate School of Management
University of Plymouth
Plymouth, Devon
UK

Festus Oderanti
Graduate School of Management
University of Plymouth
Plymouth
UK

Boris Delibašić
Faculty of Organizational Science
University of Belgrade
Belgrade
Serbia

ISSN 1865-1348   ISSN 1865-1356   (electronic)
Lecture Notes in Business Information Processing
ISBN 978-3-319-32876-8   ISBN 978-3-319-32877-5   (eBook)
DOI 10.1007/978-3-319-32877-5

Library of Congress Control Number: 2016935984

This Springer imprint is published by Springer Nature
The registered company is Springer International Publishing AG Switzerland

EURO Working Group on Decision Support Systems

The EWG-DSS is a Euro Working Group on Decision Support Systems within EURO, the Association of the European Operational Research Societies. The main purpose of the EWG-DSS is to establish a platform for encouraging state-of-the-art high-quality research and collaboration work within the DSS community. Other aims of the EWG-DSS are to:

- Encourage the exchange of information among practitioners, end-users, and researchers in the area of decision systems
- Enforce the networking among the DSS communities available and facilitate activities that are essential for the start up of international cooperation research and projects
- Facilitate the creation of professional, academic, and industrial opportunities for its members
- Favor the development of innovative models, methods, and tools in the field of decision support and related areas
- Actively promote the interest on decision systems in the scientific community by organizing dedicated workshops, seminars, mini-conferences, and conference, as well as editing special and contributed issues in relevant scientific journals

The EWG-DSS was founded with 24 members, during the EURO Summer Institute on DSS that took place at Madeira, Portugal, in May 1989, organized by two well-known academics of the OR community: Jean-Pierre Brans and José Paixão. The EWG-DSS group has substantially grown along the years. Currently, we have over 300 registered members from around the world.

Through the years, much collaboration among the group members has generated valuable contributions to the DSS field, which resulted in many journal publications. Since its creation, the EWG-DSS has held annual meetings in various European countries, and has taken active part in the EURO Conferences on decision-making-related subjects. Starting from 2015, the EWG-DSS established its own annual conferences, namely, the International Conference on Decision Support System Technology (ICDSST).

The current EWG-DSS Coordination Board comprises seven experienced scholars and practitioners in DSS field: Pascale Zaraté (France), Fátima Dargam (Austria), Rita Ribeiro (Portugal), Shaofeng Liu (UK), Boris Delibašić (Serbia), Isabelle Linden (Belgium), and Jason Papathanasiou (Greece).

The Editors:

Professor Shaofeng Liu
University of Plymouth, UK

Professor Boris Delibašić
University of Belgrade, Serbia

Dr. Festus Oderanti
University of Plymouth, UK

Preface

This sixth edition of the EWG-DSS Decision Support Systems published in the LNBIP series presents a selection of reviewed and revised papers from the Second International Conference on Decision Support System Technology (ICDSST 2016) held in Plymouth, UK, during May 23–25, 2016, with the main topic "Decision Support Systems Addressing Sustainability and Societal Challenges." This event was organized by the Euro Working Group on Decision Support Systems (EWG-DSS) in collaboration with the University of Plymouth.

The EWG-DSS series of International Conference on Decision Support System Technology (ICDSST), starting with ICDSST 2015 in Belgrade, were planned to consolidate the tradition of annual events organized by the EWG-DSS in offering a platform for European and international DSS communities, comprising the academic and industrial sectors, to present state-of-the-art DSS research and developments, to discuss current challenges that surround decision-making processes, to exchange ideas about realistic and innovative solutions, and to co-develop potential business opportunities.

The scientific topic areas of ICDSST 2016 include:

- DSS for health, demographic change, and well-being
- DSS for food security, sustainable agriculture and forestry, and bio-economy
- DSS for marine, maritime, and inland water research
- DSS for secure, clean, and efficient energy
- DSS for smart, green, and integrated transport
- DSS for sustainable construction and architecture
- DSS for climate action, environment, resource efficiency, and raw materials
- DSS for inclusive, innovative, and reflective societies
- DSS for secure societies – protecting freedom and security of the world and its citizens
- DSS for business sustainability, innovation, and entrepreneurship
- DSS for lean operations, reverse logistics, and sustainable supply chain management
- DSS for green marketing management and new product/service development
- DSS for human resource/leadership management and organizational ethics
- DSS for socially responsible accounting, finance and banking management
- DSS for economics sustainability and regional/international development
- Innovative decision-making approaches/methods and technologies
- Big data analytics for solving societal decision-making issues
- Knowledge management and business intelligence
- Decision making in modern education

This wide and rich variety of topic areas allowed us, on the one hand, to present a collection of innovative solutions to real decision-making processes in a range of domains, and, on the other hand, to highlight the main trends and research evolution in DSS. This LNBIP Springer volume edited by the EWG-DSS has considered contributions from a large number of submissions that were reviewed using a triple-blind paper evaluation method, maintaining long-established reputation and standards of high quality of the EWG-DSS. Each selected paper was reviewed by three internationally known experts from the ICDSST 2016 Program Committee comprising 79 scholars and practitioners from 28 countries. Therefore, through a rigorous two-stage, triple-blind review process, 15 out of 51 submissions were selected to be included in this volume, giving an acceptance rate of 29.41 %.

In this context, the selected papers are representative of the current and relevant DSS research and application advances. The papers have been classified into four major categories: (1) DSS applications addressing sustainability and societal challenges such as in agriculture, business ecosystem, health care, and environmental improvement; (2) DSS to support business resilience/risk management and project portfolio management; (3) DSS technologies underpinned by business intelligence and knowledge management; (4) DSS technology improving system usability and feasibility. The first category comprises five papers, the second category three papers, the third category five papers, and fourth category two papers.

We would like to thank many people who greatly helped the success of this LNBIP book. First of all, we would like to thank Springer for giving us the opportunity to guest edit the DSS book, and we especially wish to express our sincere gratitude to Ralf Gerstner and Eléonore Samklu, who have provided us with timely professional guidance and advice during the volume editing process. Secondly, we need to thank all the authors for submitting their state-of-the-art work to be considered for the LNBIP volume. All selected papers are of extremely high quality. It was a hard decision for the guest editors to select the best 15. Thirdly, we wish to express our gratitude to the reviewers, who volunteered to help with the selection and improvement of the papers.

Finally, we believe that this EWG-DSS Springer LNBIP volume has madé a high-quality selection of well-balanced and interesting research papers addressing the conference main theme. We hope the readers will enjoy the publication!

March 2016
Shaofeng Liu
Boris Delibašić
Festus Oderanti

Organization

Conference Chairs

Shaofeng Liu	University of Plymouth, UK
Boris Delibašić	University of Belgrade, Serbia
Pascale Zaraté	IRIT/Toulouse University, France
Fátima Dargam	SimTech Simulation Technology, Austria
Rita Ribeiro	UNINOVA – CA3, Portugal
Isabelle Linden	University of Namur, Belgium
Jason Papathanasiou	University of Macedonia, Greece
Jorge E. Hernández	University of Liverpool, UK

Program Committee

Abdelkader Adla	University of Oran 1, Algeria
Adiel Teixeira de Almeida	Federal University of Pernambuco, Brazil
Alan Pearman	Leeds University, UK
Alex Duffy	University of Strathclyde, UK
Alexander Smirnov	Russian Academy of Sciences, Russia
Alexis Tsoukias	University of Paris Dauphine, France
Alok Choudhary	Loughborough University, UK
Ana Paula Cabral	Federal University of Pernambuco, Brazil
Ana Respício	University of Lisbon, Portugal
Andy Wong	University of Strathclyde, UK
Antonio Rodrigues	University of Lisbon, Portugal
Ben C.K. Ngan	Pennsylvania State University, USA
Bob Young	Loughborough University, UK
Bo Zhou	Liverpool John Moores University, UK
Boris Delibašić	University of Belgrade, Serbia
Carlos Henggeler Antunes	University of Coimbra, Portugal
Csaba Csaki	University College Cork, Ireland
Daba Chowdhury	University Campus Suffolk, UK
Daniel Power	University of Northern Iowa, USA
Dobrila Petrovic	Coventry University, UK
Dragana Bečejski-Vujaklija	Serbian Society for Informatics, Serbia
Fátima Dargam	SimTech Simulation Technology/ILTC, Austria
Francesca Toni	Imperial College, UK
Festus Oderanti	University of Plymouth, UK
Francisco Antunes	Beira Interior University, Portugal
François Pinet	Cemagref/Irstea, France

Frantisek Sudzina	Aalborg University, Denmark
Gabriela Florescu	National Institute for Research and Development in Informatics, Romania
Georgiana Ifrim	University College Dublin, Ireland
Gloria Philipps-Wren	Loyola University Maryland, USA
Gregory Kersten	Concordia University, Canada
Guy Camilleri	Toulouse III University/IRIT, France
Hing Kai Chan	University of Nottingham, Ningbo Campus, UK/China
Hossam S. Ismail	Xi'an Jiaotong-Liverpool University, UK/China
Ibrahim Venkat	Universiti Sains Malaysia, Malaysia
Irène Abi-Zeid	FSA – Laval University, Canada
Isabelle Linden	University of Namur, Belgium
James Marsden	University of Connecticut, USA
Jan Mares	University of Chemical Technology, Czech Republic
Jason Papathanasiou	University of Macedonia, Greece
João Lourenço	Universidade de Lisboa, Portugal
João Paulo Costa	University of Coimbra, Portugal
Jorge Freire de Souza	Engineering University of Porto, Portugal
Jorge Hernández	University of Liverpool, UK
José Maria Moreno Jimenez	Zaragoza University, Spain
Kathrin Kirchner	Berlin School of Economics and Law, Germany
Lai Xu	Bournemouth University, UK
Leonilde Varela	University of Minho, Portugal
Lynne Butel	University of Plymouth, UK
Manuel Mora	Autonomous University of Aguascalientes, Mexico
Marc Kilgour	Wilfrid Laurier University, Canada
Marcos Borges	Universidade Federal do Rio de Janeiro, Brazil
Maria Franca Norese	Politecnico di Torino, Italy
Marko Bohanec	Jozef Stefan Institute, Slovenia
Mirjana Klajić-Borštnar	University of Maribor, Slovenia
Mirko Vujošević	University of Belgrade, Serbia
Mohamed Ghalwash	Ain Shams University, Egypt, Temple University, USA
Nadia Papamichail	University of Manchester, UK
Nikolaos Matsatsinis	Technical University of Crete, Greece
Nikolaos Ploskas	Carnegie Mellon University, USA
Nilo T. Bugtai	De La Salle University, Philippines
Pascale Zaraté	IRIT/Toulouse University, France
Pavlos Delias	Kavala Institute of Technology, Greece
Petraq Papajorgji	Canadian Institute of Technology, Albania
Phil Megicks	University of Plymouth, UK
Priscila Lima	Instituto Tercio Pacitti – UFR Federal University of Rio de Janeiro, Brazil
Rita Ribeiro	UNINOVA – CA3, Portugal
Rudolf Vetschera	University of Vienna, Austria
Sean Eom	Southeast Missouri State University, USA
Shaofeng Liu	University of Plymouth, UK

Stelios Tsafarakis	Technical University of Crete, Greece
Tina Comes	University of Agder, Norway
Uchitha Jayawickrama	Staffordshire University, UK
Vikas Kumar	Asia-Pacific Institute of Management, India
Xin James He	Fairfield University, USA
Ying Xie	Anglia Ruskin University, UK
Yong Shi	Chinese Academy of Sciences, China
Yvonne van der Toorn	Tilburg University, The Netherlands
Zoran Obradovic	Temple University, USA

Steering Committee – EWG-DSS Coordination Board

Pascale Zaraté	IRIT/Toulouse University, France
Fátima Dargam	SimTech Simulation Technology, Austria
Rita Ribeiro	UNINOVA – CA3, Portugal
Shaofeng Liu	University of Plymouth, UK
Boris Delibašić	University of Belgrade, Serbia
Isabelle Linden	University of Namur, Belgium
Jason Papathanasiou	University of Macedonia, Greece

Local Organizing Team at University of Plymouth, UK

Shaofeng Liu, Local Chair
Festus Oderanti, Lecturer
Lynne Butel, Deputy Head of School
Phil Megicks, Head of School
Femi Oyemomi, PhD scholar

Main Sponsors

Euro Working Group on Decision Support Systems (http://ewgdss.wordpress.com/)

Association of European Operational Research Societies (www.euro-online.org)

Institutional Sponsors

 **Graduate School of Management, University of Plymouth, UK** (http://www.plymouth.ac.uk/)

 **Faculty of Organisational Sciences, University of Belgrade, Serbia** (http://www.fon.bg.ac.rs/eng/)

 **SimTech Simulation Technology, Austria** (http://www.SimTechnology.com)

 ILTC - Instituto de Lógica Filosofia e Teoria da Ciência, RJ, Brazil (http://www.iltc.br)

 **University of Toulouse, France** (http://www.univ-tlse1.fr/)

 **IRIT Institut de Research en Informatique de Toulouse, France** (http://www.irit.fr/)

 **UNINOVA - CA3 - Computational Intelligence Research Group** (www.uninova.pt/ca3/)

 University of Namur, Belgium (http://www.unamur.be/)

 **Management School, University of Liverpool, UK** (http://www.liv.ac.uk/management/)

 **University of Macedonia, Department of Marketing and Operations Management, Greece** (http://www.uom.gr/index.php?newlang=eng)

Industrial Sponsors

 Springer www.springer.com

 Lumina Decision Systems (www.lumina.com)

 ExtendSim Power Tools for Simulation
(http://www.extendsim.com)

 Paramount Decisions
(https://paramountdecisions.com/)

 1000 Minds (https://www.1000minds.com/)

Professional Society Sponsor

IUFRO http://www.iufro.org/

Contents

DSS Applications Addressing
Sustainability and Societal Challenges

A Decision Support System for Multiple Criteria Alternative Ranking Using TOPSIS and VIKOR: A Case Study on Social Sustainability in Agriculture

Jason Papathanasiou[1], Nikolaos Ploskas[2(✉)], Thomas Bournaris[3], and Basil Manos[3]

[1] University of Macedonia, 156 Egnatia Street, 54006 Thessaloniki, Greece
[2] Carnegie Mellon University, 5000 Forbes Avenue, Pittsburgh, PA 15213, USA
nploskas@andrew.cmu.edu
[3] Aristotle University of Thessaloniki, University Campus,
54124 Thessaloniki, Greece

Abstract. TOPSIS and VIKOR are two well-known and widely-used multiple attribute decision making methods. Many researchers have compared the results obtained from both methods in various application domains. In this paper, we present the implementation of a web-based decision support system that incorporates TOPSIS and VIKOR and allows decision makers to compare the results obtained from both methods. Decision makers can easily upload the input data and get thorough illustrative results. Moreover, different techniques are available for each step of these methods. A real-world case study on social sustainability in agriculture is presented to highlight the key features of the implemented system. The aim of this study is to classify and rank the rural areas of Central Macedonia in Northern Greece using a set of eight social sustainability indicators.

Keywords: Multiple attribute decision making · TOPSIS · VIKOR · Decision support system · Sustainable agriculture

1 Introduction

Multi-Criteria Decision Making (MCDM) is a well-known branch of operations research that can be applied for complex decisions when a lot of criteria are involved. MCDM methods are separated into Multi-Objective Decision Making (MODM) and Multi-Attribute Decision Making (MADM) [1]. The main distinction of these groups of methods is based on the determination of the alternatives. In MODM, the alternatives are not predetermined but instead a set of objective functions is optimized subject to a set of constraints. In MADM, the alternatives are predetermined and a limited number of alternatives is to be evaluated against a set of attributes. Well-known MODM methods include bounded objective function formulation, genetic algorithms, global criterion formulation and

© Springer International Publishing Switzerland 2016
S. Liu et al. (Eds.): ICDSST 2016, LNBIP 250, pp. 3–15, 2016.
DOI: 10.1007/978-3-319-32877-5_1

goal programming, while well-known MADM methods include AHP, ELECTRE, PROMETHEE, TOPSIS and VIKOR.

Various articles have compared different MADM methods. Zanakis et al. [2] compared the performance of eight MADM methods, namely ELECTRE, MEW, SAW, TOPSIS and four versions of AHP. They found out that the final rankings of the alternatives vary across methods, especially in problems with many alternatives. Opricovic and Tzeng [3] presented a comparative analysis of TOPSIS and VIKOR in order to show their similarities and differences. The analysis revealed that TOPSIS and VIKOR use different normalization techniques and that they introduce different aggregating functions for ranking. Opricovic and Tzeng [4] compared the extended VIKOR method with ELECTRE II, PROMETHEE and TOPSIS. Ranking results were similar for ELECTRE II, PROMETHEE and VIKOR. Chu et al. [5] presented a comparison of SAW, TOPSIS and VIKOR. They found out that TOPSIS and SAW had identical rankings, while VIKOR produced different rankings. They concluded that both TOPSIS and VIKOR are suitable for assessing similar problems and provide results close to reality.

The selection of the best MADM method for a specific problem is a difficult task. There are many factors that should be considered before selecting an MADM method or a combination of MADM methods. Guitouni & Martel [6] proposed a conceptual framework for articulating tentative guidelines to choose an appropriate MADM method. Recently, Roy & Slowiński [7] presented a general framework to guide decision makers in choosing the right method for a specific problem. Other methodologies have been also proposed for the selection of the best method in specific applications [8–10].

A common problem is that different MADM methods result to different ranking results. Hence, many researchers apply different MADM methods and compare the corresponding rankings. In this paper, we present the implementation of a web-based decision support system that incorporates TOPSIS and VIKOR and allows decision makers to compare the results obtained from both methods. Decision makers can easily upload the input data and get thorough illustrative results. Different techniques are available for each step of these methods and decision makers can select them to obtain rankings according to a case's needs. Finally, a real-world case study on social sustainability in agriculture is presented to highlight the key features of the implemented system. The aim of this study is to classify and rank the rural areas of Central Macedonia in Northern Greece using a set of eight social sustainability indicators.

The remainder of this paper is organized as follow. TOPSIS and VIKOR are reviewed in Sect. 2. In Sect. 3, the implemented decision support system is presented. Section 4 presents the real-world case study on social sustainability in agriculture that have been performed to highlight the key features of the implemented system. Finally, the conclusions of this paper are outlined in Sect. 5.

2 MADM Methods

Let us assume that an MADM problem has m alternatives, $A_1, A_2, \cdots, A_m$, and n decision criteria, $C_1, C_2, \cdots, C_n$. Each alternative is evaluated with respect to the n criteria. All the alternatives' evaluations form a decision matrix $X = (x_{ij})_{m \times n}$. Let $W = (w_1, w_2, \cdots, w_n)$ be the vector of the criteria weights, where $\sum_{j=1}^{n} w_j = 1$.

This Section presents TOPSIS and VIKOR methods. Moreover, different techniques used in each step of these methods are discussed.

2.1 TOPSIS

TOPSIS (Technique of Order Preference Similarity to the Ideal Solution) method [11,12] is one of the most classical and widely-used MADM methods. TOPSIS is based in finding ideal and anti-ideal solutions and comparing the distance of each one of the alternatives to those. It has been successfully applied in various application areas, like supply chain management, logistics, engineering and environmental management [13–18].

TOPSIS method is comprised of the following five steps:

– **Step 1. Calculation of the Weighted Normalized Decision Matrix:**
  The first step is to normalize the decision matrix in order to eliminate the units of the criteria. The normalized decision matrix is computed using the vector normalization technique as follows:

$$r_{ij} = \frac{x_{ij}}{\sqrt{\sum_{i=1}^{m} x_{ij}^2}}, i = 1, 2, \cdots, m, j = 1, 2, \cdots, n$$

Another widely-used technique is the linear normalization technique. The normalized decision matrix is computed using the linear normalization technique as follows:

$$r_{ij} = \frac{x_{ij}}{x_j^+}, i = 1, 2, \cdots, m, j = 1, 2, \cdots, n, x_j^+ = max_i x_{ij}$$

for benefit criteria, and

$$r_{ij} = \frac{x_{ij}}{x_j^-}, i = 1, 2, \cdots, m, j = 1, 2, \cdots, n, x_j^- = min_i x_{ij}$$

for cost criteria. Several other normalization techniques can be incorporated at this step. Then, the normalized decision matrix is multiplied with the weight associated with each of the criteria. The normalized weighted decision matrix is calculated as follows:

$$v_{ij} = w_j r_{ij}, i = 1, 2, \cdots, m, j = 1, 2, \cdots, n$$

where w_j is the weight of the jth criterion.

- **Step 2. Determination of the Ideal and Anti-ideal Solutions:** The ideal (A^+) and anti-ideal (A^-) solutions are computed as follows:

$$A^+ = \left(v_1^+, v_2^+, \cdots, v_n^+\right) = \{(max_j v_{ij}|j \in \Omega_b), (min_j v_{ij}|j \in \Omega_c)\}, j = 1, 2, \cdots, n$$

$$A^- = \left(v_1^-, v_2^-, \cdots, v_n^-\right) = \{(min_j v_{ij}|j \in \Omega_b), (max_j v_{ij}|j \in \Omega_c)\}, j = 1, 2, \cdots, n$$

where Ω_b is the set of the benefit criteria and Ω_c is the set of the cost criteria. Another technique is to use absolute ideal and anti-ideal points, that is:

$$A^+ = (1, 1, \cdots, 1), A^- = (0, 0, \cdots, 0)$$

- **Step 3. Calculation of the Distance from the Ideal and Anti-ideal Solutions:** The distance from the ideal and the anti-ideal solutions is computed for each alternative as follows:

$$D_i^+ = \sqrt{\sum_{j=1}^{n} \left(v_{ij} - v_j^+\right)^2}, i = 1, 2, \cdots, m$$

$$D_i^- = \sqrt{\sum_{j=1}^{n} \left(v_{ij} - v_j^-\right)^2}, i = 1, 2, \cdots, m$$

Apart from the Euclidean distance, the Manhattan distance

$$D_i^+ = \sum_{j=1}^{n} \left|v_{ij} - v_j^+\right|, i = 1, 2, \cdots, m$$

$$D_i^- = \sum_{j=1}^{n} \left|v_{ij} - v_j^-\right|, i = 1, 2, \cdots, m$$

and the Chebyshev distance

$$D_i^+ = max\left(\left|v_{ij} - v_j^+\right|\right), i = 1, 2, \cdots, m$$

$$D_i^- = max\left(\left|v_{ij} - v_j^-\right|\right), i = 1, 2, \cdots, m$$

can be used.
- **Step 4. Calculation of the Relative Closeness to the Ideal Solution:** The relative closeness of each alternative to the ideal solution is calculated as follows:

$$C_i = \frac{D_i^-}{D_i^+ + D_i^-}, i = 1, 2, \cdots, m$$

where $0 \leq C_i \leq 1$.
- **Step 5. Ranking the Alternatives:** The alternatives are ranked from best (higher relative closeness value C_i) to worst.

2.2 VIKOR

VIKOR (VlseKriterijumska Optimizacija I Kompromisno Resenje) method [3] is a widely-used MADM method. The method has been developed to provide compromise solutions to discrete multiple criteria optimization problems that include conflicting criteria that usually are expressed in different units. It has been successfully applied in various application areas, like supply chain management, logistics, engineering and environmental management [19–24].

VIKOR method is comprised of the following five steps:

– **Step 1. Calculation of the Aspired and Tolerable Levels:** The first step is to determine the best f_j^+ values (aspired levels) and the worst f_j^- values (tolerable levels) of all criterion functions, $j = 1, 2, \cdots, n$:

$$f_j^+ = max_i f_{ij}, f_j^- = min_i f_{ij}, j = 1, 2, \cdots, n$$

for benefit criteria, and

$$f_j^+ = min_i f_{ij}, f_j^- = max_i f_{ij}, j = 1, 2, \cdots, n$$

for cost criteria.
– **Step 2. Determination of the Utility and the Regret Measures:** The utility measure S_i and the regret measure R_i are computed as follows:

$$S_i = \sum_{j=1}^{n} w_j (f_j^+ - f_{ij})/(f_j^+ - f_j^-), i = 1, 2, \cdots, m$$

$$R_i = max_j \left\{ w_j (f_j^+ - f_{ij})/(f_j^+ - f_j^-) \right\}, i = 1, 2, \cdots, m$$

– **Step 3. Calculation of the VIKOR Index:** The VIKOR index is computed for each alternative as follows:

$$Q_i = v \left(S_i - S^+ \right) / \left(S^- - S^+ \right) + (1 - v) \left(R_i - R^+ \right) / \left(R^- - R^+ \right), i = 1, 2, \cdots, m$$

where $S^+ = min_i S_i$, $S^- = max_i S_i$, $R^+ = min_i R_i$, $R^- = max_i R_i$; and v is the weight of the strategy of the maximum group utility (and is usually set to 0.5), whereas $1 - v$ is the weight of the individual regret.
– **Step 4. Ranking the Alternatives:** The alternatives are ranked decreasingly by the values S_i, R_i and Q_i. The results are three ranking lists.
– **Step 5. Finding a Compromise Solution:** The alternative A^1, which is the best ranked by the measure Q (minimum), is proposed as a compromise solution if the following two conditions are satisfied:
– **C1.** Acceptable advantage:

$$Q \left(A^2 \right) - Q \left(A^1 \right) \geq DQ$$

where A^2 is the second best ranked by the measure Q and $DQ = \frac{1}{m-1}$; m is the number of alternative solutions.

- **C2.** Acceptable stability in decision making: The alternative A^1 must also be the best ranked by the measures S and/or R. This compromise solution is stable within a decision making process, which could be one of the following strategies: (i) maximum group utility $(v > 0.5)$, (ii) consensus $(v \approx 0.5)$, or (iii) veto $(v < 0.5)$.

If one of the conditions is not satisfied, then a set of compromise solutions is proposed, which consists of:

- Alternatives A^1 and A^2 if only condition $C2$ is not satisfied.
- Alternatives $A^1, A^2, \cdots, A^k$ if condition $C1$ is not satisfied; A^k is determined by the relation $Q\left(A^k\right) - Q\left(A^1\right) < DQ$ for maximum k (the positions of these alternative solutions are "in closeness").

These are the steps of the original version of the VIKOR method that is used in the implemented decision support system. The method was extended at a later stage with 4 new steps which provided a stability analysis to determine the weight stability intervals and included a trade-off analysis [4, 25].

2.3 TOPSIS vs. VIKOR

A brief description of the main differences of TOPSIS and VIKOR is presented in this section. A detailed comparison of TOPSIS and VIKOR can be found in the article by Opricovic & Tzeng [3]. The main differences of these methods are the following [3]:

- **Normalization:** Both methods use a normalization technique to eliminate the units of criterion functions. The difference appears in the normalization technique used by each method. TOPSIS uses vector normalization and the normalized values depend on the evaluation unit of a criterion. On the other hand, VIKOR uses linear normalization and the normalized values do not depend on the evaluation unit of a criterion. However, a later version of TOPSIS uses linear normalization. In the proposed DSS, we provide the opportunity for the decision maker to select different normalization techniques.
- **Aggregation:** TOPSIS introduces the ranking index, including distances from the ideal and the anti-ideal point. On the other hand, VIKOR utilizes an aggregating function that represents the distance from the ideal solution. VIKOR ranking index is an aggregation of the relative importance of all criteria and a balance between the total and individual importance. TOPSIS ranking index does not include the relative importance of the ideal and anti-ideal distances; they are simply summed.
- **Solution:** Both methods provide a ranking order. The highest ranked alternative by TOPSIS is the best in terms of ranking index, which does not mean that it is always the closest to the ideal solution. On the other hand, the highest ranked alternative by VIKOR is always the closest to the ideal solution. Moreover, VIKOR proposes a compromise solution with an advantage rate.

3   Implementation and Presentation of the Decision Support System

The web-based decision support system has been implemented using PHP, MySQL, Ajax and jQuery. Initially, the decision maker should upload the data of the case study and adjust methods' parameters (Fig. 1). Decision makers can download an Excel template, incorporate their data and upload the Excel file to the decision support system. Moreover, they can select different parameters for each method. More specifically, decision makers can select the normalization technique (vector or linear), the technique to calculate the ideal and anti-ideal solutions (min/max or absolute values) and the distance measure to be used (Euclidean, Manhattan or Chebyshev) for TOPSIS method and the weight of the maximum group utility strategy (v) for VIKOR method. The results are graphically and numerically displayed, allowing the decision makers to easily compare the rankings obtained by the two methods (Fig. 2). The DSS can also output a thorough report in a pdf file containing the results of TOPSIS and VIKOR. The result of TOPSIS is the ranking index, while the result of VIKOR is a compromise solution (if the acceptable advantage condition (C1) and the acceptable stability condition (C2) are met) or a set of compromise solutions.

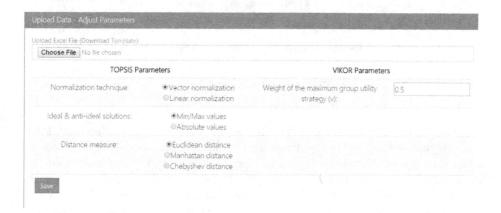

Fig. 1. Upload data & Adjust parameters

4   Case Study

The aim of this case study is to classify and rank the rural areas of Central Macedonia in Northern Greece using a set of eight social sustainability indicators. In order to measure these indicators, a survey was conducted in farm households of the Region of Central Macedonia in Northern Greece. The sample of the survey was 145 farm households from the seven prefectures of the region (Chalkidiki,

Fig. 2. Results & Ranking

Imathia, Kilkis, Pella, Pieria, Serres, Thessaloniki) who have subsidized with direct payments from the first pillar of the Common Agricultural Policy. The aim of the survey was to measure the social sustainability of the farm households in European Union rural areas. The survey included a detailed questionnaire with personal and phone interviews. A large number of social and economic indicators was measured. From this set of indicators, we have selected 8 indicators that can represent the main social characteristics of the farm households. The selected indicators are the following:

1. **Highest Education Level Attained by One Household's Member:** According to OECD [26]: "Education plays a key role in providing individuals with the knowledge, skills and competences needed to participate effectively in society and in the economy". Hence, this is a benefit criterion (the highest education level of at least one member of the farm household would increase the farmer's knowledge and skills). The scale used for this criterion is the following: 1 - elementary, 2 - primary, 3 - high school, 4 - bachelor, 5 - master, and 6 - PhD.

2. **Number of Employed in the Farm Household:** According to Eurostat Labour Force Survey (LFS) [27]: an employed person is "a person aged 15 and over who during the reference week performed work - even if just for one hour a week - for pay, profit or family gain". This is a benefit criterion.

3. **Number of Long-Term Unemployed in the Farm Household:** According to OECD [26]: "Long-term unemployment is defined as referring to people who have been unemployed for 12 months or more". This is a cost criterion.

4. **Percentage of the Total Household Gross Revenue Comes from Farming:** The gross revenue comes from farming refers to monetary and non-monetary income received by farm operators. This is a benefit criterion (the maximization of the gross revenue comes from farming would support professional farmers). The scale used for this criterion is the following: 1 - less than 10%, 2 - $10 - 29\%$, 3 - $30 - 49\%$, 4 - $50 - 69\%$, 5 - $70 - 89\%$, 6 - more than 89%.

5. **Employment Rate Percentage in the Farm Household:** According to OECD [26]: "Employment rate is defined as the proportion of working age adults employed with working age between 15 and 64 years old". This is a benefit criterion.

6. **Share of Labour Used in Off Farm Activities:** This criterion refers to the portion of the farm household income obtained by nonfarm wages and salaries, pensions, and interest income earned by farm families. This is a cost criterion (the minimization of the labour's share in off farm activities would support professional farmers).

7. **Share of the Farm Income Comes from Subsidies:** Farm subsidy is a governmental subsidy paid to farmers to support their income. This is a cost criterion (the minimization of the farm income comes from subsidies would support professional farmers).

8. **Number of Household's Members That Have a Formal Agricultural Education:** This is a benefit criterion (the formal agricultural education of at least one member of the farm household would increase the farmer's knowledge and skills).

Table 1 presents the average indicators of the data collected for each prefecture. TOPSIS method was performed using the vector normalization and the finding of the best and worst performance for the ideal and anti-ideal solutions, while VIKOR method was performed setting the weight of the strategy of the maximum group utility v equal to 0.5. The criteria are equally important ($w_j = 0.125, j = 1, 2, \cdots, n$). Table 2 and Fig. 3 present the rankings obtained from each method. TOPSIS ranks the prefecture of Kilkis as the best and the prefecture of Pieria as the worst rural area of Northern Greece, while VIKOR ranks the prefecture of Pella as the best and the prefecture of Chalkidiki as the worst rural area of Northern Greece. The rankings are not similar as TOPSIS and VIKOR use different kinds of normalization to eliminate the units of criterion functions and different aggregating functions for ranking [3].

Table 1. The decision matrix

Prefecture	Highest education level attained by one household's member	Number of employed in the farm household	Number of long-term unemployed in the farm household	Percentage of the total household gross revenue comes from farming	Employment rate percentage in the farm household	Share of labour used in off farm activities	Share of the farm income comes from subsidies	Number of household's members that have a formal agricultural education
Chalkidiki	3.70	3.30	0.35	4.95	0.50	0.03	0.48	0.30
Imathia	3.25	2.95	0.00	5.40	0.72	0.03	0.45	0.30
Kilkis	3.65	2.94	0.12	5.12	0.62	0.02	0.44	0.53
Pella	3.75	3.63	0.19	5.19	0.55	0.00	0.44	0.38
Pieria	4.19	4.06	0.50	5.19	0.50	0.15	0.43	0.44
Serres	3.71	3.18	0.24	5.06	0.55	0.07	0.46	0.26
Thessaloniki	3.55	3.14	0.00	5.09	0.59	0.01	0.53	0.18

Table 2. Results & Ranking

		Alternatives							Ranking
		Chalkidiki (A1)	Imathia (A2)	Kilkis (A3)	Pella (A4)	Pieria (A5)	Serres (A6)	Thessaloniki (A7)	
TOPSIS	D^+	0.0767	0.0423	0.0328	0.0429	0.1430	0.0786	0.0517	A3, A2, A4, A7, A1, A6, A5
	D^-	0.0929	0.1278	0.1263	0.1261	0.0407	0.0760	0.1363	A7, A2, A3, A4, A1, A6, A5
	C	0.5476	0.7514	0.7938	0.7461	0.2217	0.4919	0.7250	A3, A2, A4, A7, A1, A6, A5
VIKOR	S	0.6570	0.3810	0.3910	0.3750	0.4650	0.6050	0.6060	A4, A2, A3, A5, A6, A7, A1
	R	0.1250	0.1250	0.1250	0.0966	0.1250	0.0982	0.1250	A4, A6, A1 ≈ A2 ≈ A3 ≈ A5 ≈ A7
	Q	1.0000	0.5106	0.5284	0.0000	0.6596	0.4360	0.9096	A4, A6, A2, A3, A5, A7, A1

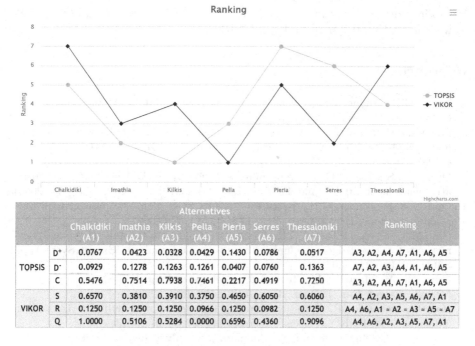

Fig. 3. Results & Ranking

		Alternatives							Ranking
		Chalkidiki (A1)	Imathia (A2)	Kilkis (A3)	Pella (A4)	Pieria (A5)	Serres (A6)	Thessaloniki (A7)	
TOPSIS	D⁺	0.0767	0.0423	0.0328	0.0429	0.1430	0.0786	0.0517	A3, A2, A4, A7, A1, A6, A5
	D⁻	0.0929	0.1278	0.1263	0.1261	0.0407	0.0760	0.1363	A7, A2, A3, A4, A1, A6, A5
	C	0.5476	0.7514	0.7938	0.7461	0.2217	0.4919	0.7250	A3, A2, A4, A7, A1, A6, A5
VIKOR	S	0.6570	0.3810	0.3910	0.3750	0.4650	0.6050	0.6060	A4, A2, A3, A5, A6, A7, A1
	R	0.1250	0.1250	0.1250	0.0966	0.1250	0.0982	0.1250	A4, A6, A1 = A2 = A3 = A5 = A7
	Q	1.0000	0.5106	0.5284	0.0000	0.6596	0.4360	0.9096	A4, A6, A2, A3, A5, A7, A1

5   Conclusions

A common problem researchers encounter when setting up comparisons of different MADM methods is that each method can result to different ranking results. Hence, many researchers apply different MADM methods and compare the corresponding rankings. In this paper, we presented the implementation of a web-based decision support system that incorporates TOPSIS and VIKOR and allows decision makers to compare the results and rankings obtained from both methods. Different techniques are available for each step of these methods. More specifically, decision makers can select the normalization technique (vector or linear), the technique to calculate the ideal and anti-ideal solutions (min/max or absolute values) and the distance measure to be used (Euclidean, Manhattan or Chebyshev) for TOPSIS method and the weight of the maximum group utility strategy (v) for VIKOR method. The results are graphically and numerically displayed, allowing the decision makers to easily compare the rankings obtained by the two methods. Finally, a real-world case study on social sustainability in agriculture was presented. The aim of this study is to classify and rank the rural areas of Central Macedonia in Northern Greece using a set of eight social sustainability indicators. Using the implemented decision support system, decision makers can easily obtain rankings by TOPSIS and VIKOR. In future work, we plan to include fuzzy versions of TOPSIS and VIKOR as well as other MADM methods like PROMETHEE and ELECTRE.

References

1. Tzeng, G.H., Huang, J.J.: Multiple Attribute Decision Making: Methods and Applications. CRC Press, Boca Raton (2011)
2. Zanakis, S.H., Solomon, A., Wishart, N., Dublish, S.: Multi-attribute decision making: a simulation comparison of select methods. Eur. J. Oper. Res. **107**(3), 507–529 (1998)
3. Opricovic, S., Tzeng, G.H.: Compromise solution by MCDM methods: a comparative analysis of VIKOR and TOPSIS. Eur. J. Oper. Res. **156**(2), 445–455 (2004)
4. Opricovic, S., Tzeng, G.H.: Extended VIKOR method in comparison with outranking methods. Eur. J. Oper. Res. **178**(2), 514–529 (2007)
5. Chu, M.T., Shyu, J., Tzeng, G.H., Khosla, R.: Comparison among three analytical methods for knowledge communities group-decision analysis. Expert Syst. Appl. **33**(4), 1011–1024 (2007)
6. Guitouni, A., Martel, J.M.: Tentative guidelines to help choosing an appropriate MCDA method. Eur. J. Oper. Res. **109**(2), 501–521 (1998)
7. Roy, B., Slowiński, R.: Questions guiding the choice of a multicriteria decision aiding method. EURO J. Decis. Processes **1**(1–2), 69–97 (2013)
8. Özcan, T., Çelebi, N., Esnaf, Ş.: Comparative analysis of multi-criteria decision making methodologies and implementation of a warehouse location selection problem. Expert Syst. Appl. **38**(8), 9773–9779 (2011)
9. Kurka, T., Blackwood, D.: Selection of MCA methods to support decision making for renewable energy developments. Renew. Sustain. Energy Rev. **27**, 225–233 (2013)
10. Mendoza, G.A., Martins, H.: Multi-criteria decision analysis in natural resource management: a critical review of methods and new modelling paradigms. For. Ecol. Manag. **230**(1), 1–22 (2006)
11. Yoon, K.P.: System Selection by Multiple Attribute Decision Making. Ph.D. Dissertation, Kansas State University, Manhattan, KS (1980)
12. Hwang, C.L., Yoon, K.: Multiple Attribute Decision Making - Methods and Applications. Springer, Heidelberg (1981)
13. Behzadian, M., Otaghsara, S.K., Yazdani, M., Ignatius, J.: A state-of the-art survey of TOPSIS applications. Expert Syst. Appl. **39**(17), 13051–13069 (2012)
14. Khorshidi, R., Hassani, A.: Comparative analysis between TOPSIS and PSI methods of materials selection to achieve a desirable combination of strength and workability in Al/SiC composite. Mater. Des. **52**, 999–1010 (2013)
15. Alimoradi, A., Yussuf, R.M., Zulkifli, N.: A hybrid model for remanufacturing facility location problem in a closed-loop supply chain. Int. J. Sustain. Eng. **4**(1), 16–23 (2011)
16. Krohling, R.A., Campanharo, V.C.: Fuzzy TOPSIS for group decision making: a case study for accidents with oil spill in the sea. Expert Syst. Appl. **38**(4), 4190–4197 (2011)
17. Liao, C.N., Kao, H.P.: An integrated fuzzy TOPSIS and MCGP approach to supplier selection in supply chain management. Expert Syst. Appl. **38**(9), 10803–10811 (2011)

18. Zavadskas, E.K., Antucheviciene, J.: Development of an indicator model and ranking of sustainable revitalization alternatives of derelict property: a Lithuanian case study. Sustain. Dev. **14**(5), 287–299 (2006)
19. Hsu, C.H., Wang, F.K., Tzeng, G.H.: The best vendor selection for conducting the recycled material based on a hybrid MCDM model combining DANP with VIKOR. Resour. Conserv. Recycl. **66**, 95–111 (2012)
20. Shemshadi, A., Shirazi, H., Toreihi, M., Tarokh, M.J.: A fuzzy VIKOR method for supplier selection based on entropy measure for objective weighting. Expert Syst. Appl. **38**(10), 12160–12167 (2011)
21. Caterino, N., Iervolino, I., Manfredi, G., Cosenza, E.: Comparative analysis of multi-criteria decision-making methods for seismic structural retrofitting. Comput. Aided Civ. Infrastruct. Eng. **24**(6), 432–445 (2009)
22. Tzeng, G.H., Huang, C.Y.: Combined DEMATEL technique with hybrid MCDM methods for creating the aspired intelligent global manufacturing & logistics systems. Ann. Oper. Res. **197**(1), 159–190 (2012)
23. Chang, C.L., Hsu, C.H.: Multi-criteria analysis via the VIKOR method for prioritizing land-use restraint strategies in the Tseng-Wen reservoir watershed. J. Environ. Manag. **90**(11), 3226–3230 (2009)
24. Opricovic, S.: Fuzzy VIKOR with an application to water resources planning. Expert Syst. Appl. **38**(10), 12983–12990 (2011)
25. Opricovic, S.: A compromise solution in water resources planning. Water Resour. Manag. **23**(8), 1549–1561 (2009)
26. OECD: OECD Factbook: Economic Environmental and Social Statistics. OECD Publishing (2014)
27. EU LFS: Eurostat Labour Force Survey Glossary. Online Publication (2014)

An Operations Research-Based Morphological Analysis to Support Environmental Management Decision-Making

Maria de Fátima Teles[1](✉) and Jorge Freire de Sousa[2]

[1] CP-Comboios de Portugal, E.P.E.,
Praça Almeida Garrett, Estação Porto São Bento,
4000-069 Porto, Portugal
mariafatimateles@gmail.com
[2] FEUP and CEGI – INESC TEC,
Rua Dr. Roberto Frias, 4200-465 Porto, Portugal

Abstract. In this paper the authors present a meta-model aiming to support decision-makers that wish to know more about how to use systems models to cope with the integration of environmental concerns into the company strategy. This is made by using a General Morphological Analysis (GMA) to bridge the gap between Operations Research (OR) analysts, decision-makers and stakeholders, making all of them part of the problem structuring and formulation process, particularly in societal issues like the environmental ones. The novelty of this approach is two-fold: (i) there are no examples in literature of a GMA research that address a linkage between environmental practices, strategic objectives, and the integration of stakeholders in the decision-making process at the level of a company; (ii) there is no GMA that had covered all the phases of a decision-making problem (problem definition, problem analysis and problem solving) in such a context.

Keywords: General morphological analysis · Environmental management · Strategic decision-making · Stakeholder involvement

1 Introduction

1.1 Context and Motivation

Some of the main social concerns of humanity (such as economic growth and environmental protection) are now at the core of the strategy and decision-making mechanisms of the companies, demanding more and more stakeholder involvement. As a consequence, current and future managers are obliged to call upon a great variety of models to deal with decision-making in order to gain a sustainable and more competitive position, imposing the adoption of new "methods for structuring policy spaces and strategy alternatives, and for organizational planning" [1].

OR Agent: An Analyst and a Facilitator. OR models are increasingly being relied upon to inform and support environmental management. "They are incorporating an even broader range of disciplines and now often confront people without strong

© Springer International Publishing Switzerland 2016
S. Liu et al. (Eds.): ICDSST 2016, LNBIP 250, pp. 16–30, 2016.
DOI: 10.1007/978-3-319-32877-5_2

quantitative or model-building backgrounds. These trends imply a need for wider awareness of what constitutes good model-development practice" [2]. When dealing with problem situations at a more strategic level, there may be several reasons to the operational researcher acting not only as an analyst but also as a facilitator to the client: lack of agreement on the scope of the problem to be addressed; the existence of several stakeholders with distinct perspectives, objectives, values and interests; and the variable levels of participation required in the decision making process. All these factors may have a significant influence on making the solutions derived from the analysis not only desirable for the client but politically feasible to the organization [3]. Additionally, many of the factors involved are not quantifiable, since they contain strong socio-political dimensions and conscious self-reference among actors. This means that traditional quantitative methods, causal modeling and simulation alone may be relatively useless [4].

Behavior and Ethics: New Modeling Drivers. Behavioral elements become ever more important as we move from optimization to solving people related problems. Stakeholders have not only different perspectives but different mental models, and creating a common language is a key step, particularly in environmental issues [5]. "Maintaining credibility and 'staying onside' with different stakeholders might require the analyst to 'tread carefully' and negotiate their way through difficult political situations; making informed decisions about what method or combination of methods might best 'fit' in a particular intervention might require as much consideration be given to the receptiveness of the people involved to different approaches as it does to the perceived needs of the situation under investigation" [6]. Models are being used to solve and to help understand complex environmental problems. "Modelers with high ethical standards must be open to acknowledge the risks of behavioral effects" [7]. Some biases can be derived from involuntary cognitive limitations but others can be strategically motivated omissions or over or under emphasis of some features. Recent global events such as the threat of environmental disaster, corporate dishonesty, and the credit crunch caused a greater public recognition of the ethical implications of management actions [8].

1.2   Scope of the Research

The work described in this paper is part of a broader research project aiming (i) to provide a methodology or framework to support and evaluate corporate environmental strategies and management approaches; (ii) to propose a methodological approach for structuring multi-objective problems involving multi-stakeholder decision-making in a participatory context, searching for value co-creation from both societal and corporate perspectives; (iii) to apply the theoretical conclusions drawn from the previous objectives to a case study in the context of a bus public transport company [9]. In order to deal with a variety of pertinent issues identified in different approaches – though at the organizational level – we developed a framework aiming to promote the integration of environmental policies in the strategy of a company [10]. But there was a need of a methodology for structuring problems involving multiple stakeholders in a participatory decision-making context. In the path taken along this 'journey', it became increasingly clear that finding a way of classifying and comparing different types of modeling methods employed in the OR and Management Science (MS) field on the basis of a number of contextual properties would augment the likelihood of success in their application. We will discuss now a few reasons for the plausibility of this belief.

OR Interventions: A Mean to Transfer Knowledge to Companies. There are plenty of articles on problem structuring methods but very few focus them as interventions, meaning systematic or purposeful actions by an agent to create change or improvements. These interventions lead to "complex connections between various actors (human and non-human), pursuing their personal interest, in a flux of changing circumstances and context" [11]. And these questions relate to the role of values in socio-environmental modeling. As stated by Hämäläinen, "applied science is not value free; value dependence is a strong driver of behavioral effects" [7]. OR interventions intend to facilitate thinking and problem solving. So, a balance between models and people skills is needed. Developing facilitation and systems intelligence skills is a must in the action of an OR analyst. As "there are usually multiple paths that can be followed in a decision analysis process" and "it is possible that these different paths lead to different outcomes" [12], unintentional biases in model use may occur. The importance of this aspect is reinforced by the fact that many non-experts in companies are increasingly seduced by apparently easy-to-use technical software, ever more pervasive, making them unconscious preys of pitfalls and risks.

One of our primordial purposes is indeed to promote and enhance the transference of knowledge to and within the companies. We see companies as living 'cells' in the economic and social structure where specific characteristics must be present in order to assure resilience. "Change (disturbance), learning, diversity, and self-organization" are critical attributes of resilient systems [13]. The absence of a holistic and integrated 'toolbox' for company decision-makers in the context mentioned above revealed that there was room to construct a typology of modeling methods. And, with General Morphological Analysis, from now on GMA [14, 15] there was an opportunity to explicitly materialize and propose a possible typology of decision-making modeling methods to approach in different ways company's problems in hand. GMA can be defined as "a method for structuring and investigating the total set of relationships contained in multi-dimensional, non-quantifiable, problem complexes" [16].

GMA: A Bridge Between Academia and Companies. Müller-Merbach was the author of one of the earliest appeals to use GMA in OR programs [17], pointing out "that general morphology is especially suitable for OR, not the least because of the growing need for operational analysts to be part of the problem structuring and formulation process, and not simply a 'receiver' of predefined problems" [18]. Our intention is not to provide the design of a GMA but rather to make available a meta-model based on OR in the environmental field for current and future managers of companies. This is particularly important if we are taking into account the current needs of establishing bridges between High Education and managing companies in a sustainable way [19]. The methods included in our GMA "are not prescriptive, in that it will never tell us how to conduct an intervention. They are tools for reflection and learning, a structure and method to interrogate practice so that learning takes place;

they do not necessary lead to rules for action. To understand complex interventions, a complex of approaches is necessary" [11].

Structure of the Paper. The remaining of the paper is organized as follows. A brief literature review on GMA applications to business is presented in Sect. 2. Section 3 refers the way in which the GMA was developed. In Sect. 4, all the parameters used are identified and justified. The Cross-Consistency Assessment is the object of Sect. 5. Finally, in Sect. 6, we present the conclusions of the undertaken work.

2   Business Applications of GMA - A Brief Literature Review

In a recent article, Álvarez and Ritchey [1] present 80 examples of applications of Morphological Analysis (MA) since the 1950's until the present. The applications overviewed cover categorized areas, such as: engineering and product design; general design theory and architecture; futures studies and scenario development; technological forecasting; management science, policy analysis and organizational design; security, safety and defense studies; creativity, innovation and knowledge management and modeling theory, OR methods and GMA itself.

The latest articles concerned with business models and management in companies are here reviewed in order to focus their relevance for our field of research.

MA is used as a tool to aid both in strategy design and development within the organization. In Swanich [20], for instance, it is employed for strategic planning of a business company regarding regulatory uncertainty in the airline industry. An approach concerning new business model development is proposed by Im and Cho [21]. The authors aim "to support practitioners to develop, evaluate and select the best business model to achieve the business objectives". The first stage of the approach uses MA to derive and aggregate a set of possible business model alternatives. And, in the second stage, the alternatives are evaluated and selected through the adoption of fuzzy extent analytic hierarchy process (FAHP) and fuzzy techniques for order of preference by similarity to ideal solution (TOPSIS). Other authors [22] highlight the importance of using MA as a "superior method" to generate and combine business model alternatives. MA is also employed to generate or develop, in a systematic and structured way, new business models centered on strategic marketing and innovation management, based on a methodology which covers several phases of development of ideas and prototyping. Another research [23] presents a strategic approach to identify the possible options of co-creation design. These authors develop a framework that integrates the multiple design dimensions, reveals new co-creation opportunities, and explores how the company may identify them with the use of MA. A different study links business market with MA [24]. The authors apply MA in a competitive arena mapping procedure, which enables firms to systematically plot possible competitive arenas and use managerial judgment to select those which are growing and for which the firm has exploitation capabilities. In Heorhiadi et al., the management of the organizational structure of a company is considered as a "multilevel system organized according to functional definitions of its structural components". These authors consider the use of

MA as an aid to possible identification of "the causes of problems emerging within the existing management organizational structures" [25]. MA has also been applied to the design of a production process by [26] in a two-step approach: beginning with the MA procedure (specification of the production system classes) and then generating variants examples of existing type of production systems.

The novelty of our approach is two-fold. On one side, there are no examples in literature of an MA research that address a linkage between environmental practices, strategic objectives, and the integration of stakeholders in the decision-making process at the level of a company. On other side, to the authors' knowledge, there is no MA that had covered all the phases of a decision-making problem (problem definition, problem analysis and problem solving) in such a context, as we are doing.

3   The Development of the GMA

The development of our GMA relied primarily on the form and identification of the diverse dimensions that compound the conceptual (organizational) system as a whole [18]. We guided our analysis in a systematic way by establishing a 'bridge' among three broad guiding vectors: *Context, Stakeholders* and *OR Methodologies*. For each of these guiding vectors, we identified and defined the most important dimensions (or parameters). An initial Morphological Model began to take shape comprising fields of constructed dimensions (morphological fields).

To undertake this task in such a system requires different levels of abstraction and thus, in a first moment, we developed a GMA using 'linking fields' in the sense proposed by [18]:"it is possible to allow the designated output of one morphological field to become the input for another field. Alternatively, the designated output a number of (sub-) fields can be collated into a single (super-) field. This allows for a hierarchical or networked morphological model." We use the term "sub-model" as a synonym of "linking fields". We have defined the context of the GMA as the Decision-Making Process for Designing (or Redesigning) the Strategy of an Organi-zation System that pretends to address (and integrate) environmental concerns. In the *Context*, we have constructed two sub-models: one related with *Stakeholders* and other grounded on OR science designated *Systemic Methodologies*. The *Context* and *Stakeholders* guiding vectors have, respectively, five and four dimensions, while *Systemic Methodologies* presents three dimensions. For each dimension we assigned a range of relevant values or conditions in a natural language.

Despite this first development of the GMA, we determined the need to use a more "manageable" model and therefore we developed some simplifications, not supported on hierarchical and linked models, resulting in a six parameter model to carry out.

The next section presents the fundamental variables identified throughout the stages of constructing the GMA in both versions. The identification of these fundamental variables is supported by references of specialized literature.

4 Identification of the Parameters

In 1984, Jackson and Keys pointed out the need to exist a "coordinated research programme designated to deepen understanding of different problem context and the type of problem-solving methodology appropriate to each" [27]. These authors created a framework named System of Systems Methodologies which had in its roots a classification of systems based on an "ideal-type" grid for problem situations. Later on, Jackson presented an extended version of the "ideal-type" grid which contained two dimensions in its axes: "participants" and "systems" [28]. The first version of our GMA lies on these works and considers three broad dimensions: *Context*, *Stakeholders* and *Systemic Methodologies*. We will describe it now, and then present the matrix representation of the parameters, their range values and their connections.

The dimension *Context* regards a decision-making process for designing (or redesigning) the strategy of an organization and has five parameters. Two of them address environmental concerns: (i) the purpose of the problem: the integration of environmental management decisions and (ii) their existing dominant paradigms.

These two parameters are supported on the work provided by Gregory et al. (2012) as these authors cover this field of study in a holistic way [29]. The parameters' head are based on this research. However, their single values were adapted. The authors consider science-based decision-making, consensus-based decision-making and analyses based in economics or multi-criteria decision techniques (MCDT) as the three guides to environmental decisions. We considered more adequate to our purpose to maintain the first one and to adopt stakeholders-based (rather than consensus-based) decision-making due to the relevance of stakeholders and their judgmental processes, and also because consensus may not always be a goal, mainly in companies. The third paradigm was changed to "economic-based decision-making" because, together with the other two, it provides a connection with the three basic pillars of corporate sustainability. Besides, in our research, MCDT are conceptually placed in the *Systemic Methodologies* vector.

At this point, we have already defined what is going to be modeled and the purpose of that modeling. But another parameter should be reflected in the model: the desired final result. Their range values were "borrowed" from [30]. The two other parameters under the 'umbrella' of *Context* are related with "who is part of the decision-making process" and with the "nature of the problem". In these two parameters we considered possible redirections to sub-models *Stakeholders* and *Systemic Methodologies*, respectively. In other words, we propose a GMA with linking fields (see Sect. 3).

The first one considers three types of situations, regarding the participants in the decision-making process: only the decision-maker (DM), the DM along with some stakeholders or the involvement of all the stakeholders. If one of these two last options is selected this output will constitute the input for sub-model *Stakeholders*.

The last parameter in *Context* is related to "the nature of the problem". The problems may be classified in two broad types: objective, well-defined problems or subjective, ill-defined ones. The first type has usually an uncontested formulation, susceptible of technical solutions and that does not require the inclusion of the subjective opinion of individuals. The second type of problems usually needs to include different viewpoints of participants to be defined. Even if we recognize several variants to this classification, for simplification we have just considered two parameter-values: "simple/objective/well-defined/well-structured" problems and "complex/subjective/ill-defined/wicked" ones. We allow this simplification because the "problem nature" parameter is going to be scrutinized in two ways: as a linking field to sub-model *Systemic Methodologies*, that embodies an array of OR models; and as a consequent of another linking field: sub-model *Stakeholders*.

The path that will result from the definition of the problem context and the stakeholders' involvement is a determining factor to provide the necessary inputs to explore and allow the choice of the appropriate OR approach. In Table 1 we synthesize the dimension *Context*.

Table 1. *Context* of the organizational system

Context: a Decision-Making Process for Designing (or Redesigning) the Strategy of an Organizational System				
Purpose: Integration of Environmental Management Decisions [IEMD]	Desired final result	Environmental Management Decisions [EMD] (Dominant Paradigms)	Who is part of the Decision-Making Process [DMP]	Nature of the Problem (then go to Sub-model "Systemic Methodologies")
Choosing a single preferred alternative (a solution to a policy/planning problem)	To better structure and define the problem	Science-based decision making	Only the owner/client of the problem (DM)	Simple/Objective/Well defined/Well Structured
Developing a system for repeated choices (decisions that are likely to be repeated)	To establish and legitimate an idea or policy direction	Stakeholders based decision making	The DM and other stakeholders (then go to Sub-model "Stakeholders")	
Making linked choices (a way to separate decisions to be made now as opposed to later)	Specific proposal for solution			Complex/Subjective/Ill-defined/"Wicked"
Ranking (a way to put actions in order of importance or preference)	Increase knowledge and competence within problem area	Economic-based decision making	All the stakeholders are involved (then go to Sub-model "Stakeholders")	
Routing (grouping of actions into different categories)				

Sub-model *Stakeholders* contains four parameters, as shown in Table 2. The first one regards the level of participation of the stakeholders, going from *just being kept informed* to *being implemented the option according to what stakeholders did decide*. We adopt this categorization in the terms proposed by [31]. Two of the other

Table 2. Sub-model *Stakeholders*

Sub-model: Stakeholders			
Levels of Participation of Stakeholders	The level of access to objective variables and data	The level of Stakeholders´ trust in the scientific methodologies for solving organizational problems	Stakeholders' diversity of Views/Values/Interests (Stakeholder VVI) (then go to Sub-model "Systemic Methodologies")
Inform (promise: we will keep you informed)	Reliable data exist and are generally available to the OR experts	Do not believe in scientific methodologies	Unitary (they share common purposes and are all involved)
Consult (promise: we will keep you informed, listen to you, and provide feedback on how your input influenced the decision)	Reliable data exist but are not generally available (epistemic uncertainty)	Believe only in the efficiency of mathematical modeling methods	Pluralistic (their basic interests are compatible but they do not share the same values and beliefs; compromises are possible)
Involve (promise: we will work with you to ensure your concerns are considered and reflected in the alternatives considered, and provide feedback on how your input influenced the decision)	Reliable data does not exist (genuine uncertainty)	Believe in the efficiency of both quantitative and qualitative modeling	Conflicting/Coercive (they have few interests in common, they are free to express them and their conflicting values do not allow a convergence towards agreed objectives)
Collaborate (promise: we will incorporate your advice and recommendations to the maximum extent)			
Empower (promise: we will implement what you decide)			

parameters are related with the reliability of data and its availability, and with the level of stakeholders' trust in scientific methods for solving organizational problems.

These two parameters as well as their values were presented in a multi-methodology model that addresses a GMA for stakeholders [30]. It is important to emphasize that this approach includes local stakeholders or individuals and that they may be incorporated in a decision-making process. They may also be engaged in gathering data from their local livelihoods, providing a precious input to modeling (e.g. in environmental issues or in transport or urban planning). Stakeholders do not need to be familiar with modeling or to believe in scientific models to be called to be an active part in the process of decision-making. The last parameter reflects stakeholders' diversity of views, values and interests. Its singular values are: unitary, pluralistic and conflicting or coercive [28, 32, 33].

At this point, and according to results obtained, different methodologies and techniques may support our problem. Thus, another level of abstraction is required: sub-model *Systemic Methodologies*. This sub-model has three parameters: methodological orientation, systems methodologies (SM) and multi-criteria decision analysis (MCDA) categories, as shown in Table 3. Our primary sources were Checkland [34] and Schwenk [35]. Since then, several authors used Systems Thinking approaches [28, 32, 33, 36–38] to map the development of management sciences along time or to classify OR/MS methodologies based on different number and type of axes.

Table 3. Sub-model *Systemic Methodologies*

Sub-model: Systemic Methodologies		
Methodological Orientation [Methodolog. Orient.]	Systems Methodologies	Multi-Criteria Analysis (MCDA) Categories
Functionalist Systems Approaches	Hard methods: Classical OR/MS techniques; Systems Dynamics; Systems Analysis Viable Systems Modeling Other Techniques	"First Generation" MCDA Methods: MAUT/MAVT; Outranking (Promethee, Electre); Pairwise Comparisons (AHP/ANP); Others (TOPSIS)
Interpretative Systems Approaches	Problem Structuring Methods: Interactive Management Robustness Analysis Soft Systems Methodology Strategic Assumption Surfacing and Testing Strategic Choice Approach Cognitive and Causal Mapping / Strategic Options Development and Analysis Drama Theory	"Second Generation" MCDA Methods: Social Multi-Criteria Analysis (SMCA); Multiple Criteria Group Decision Making (MGDM); Multi Actor Multi-criteria Analysis (MAMCA); Others
Emancipatory Systems Approaches	Critical Systems Heuristics	
Critical Systems Thinking and Multi-methodology	Total Systems Intervention meta-methodology	

In organizations, systems consist of people, structures, and processes. In our approach, when we refer to *Systemic Methodologies*, we are considering not only the SM themselves, as an array of methodologies to support Decision Analysis, but simultaneously two complementary levels of analysis: at a higher level, the methodological orientation for the SM, and, at a lower level, if, by using SM, the researchers and decision-makers take into account, in an explicit way, the opinion of the stakeholders, leading us to MCDA. The first value for this parameter, regarding MCDA categories, is entitled "First Generation" of MCDA Methods, because they do not need to internally integrate explicitly stakeholders' opinions. In contrast, MCDA Methods that need to integrate stakeholders' opinions are here called the "Second Generation". These methods may be found, respectively, in [39] and [40].

However, it would not be easy to use this model in practice, given its complexity and the number of combinations considered. A more 'manageable' model was then developed, giving place to a GMA without an explicit hierarchy or linking fields. This work assumed some simplifications, resulting in a problem space consisting in a six parameter model. Five of the parameters previously analyzed were maintained. The new parameter is the result of the fusion of two parameters presented above: one from sub-model *Context*, the parameter related to "Who is part of the Decision-Making Process" and another one from sub-model *Stakeholders*, the parameter designated by "Levels of Participation of Stakeholders". This ensemble constitutes the final version, presented in Table 4.

Table 4. The final version of the GMA

Context: a Decision-Making Process for Designing (or Redesigning) the Strategy of an Organizational System					
Purpose: IEMD	Dominant Paradigms: EMD	Who is part of the DMP	Nature of the Problem	Stakeholder VVI	Methodolog. Orient.
Choosing a single preferred alternative	Science-based decision making	Only the owner of the problem, stakeholders eventually just informed or consulted	Simple/ Objective/ Well Defined	Unitary	Functionalist Systems Approaches
Developing a system for repeated choices	Stakeholders based decision making			Pluralistic	Interpretative Systems Approaches
Making linked choices		The owner of the problem and other stakeholders involved in the DMP	Complex/ Objective/ Ill-defined/ "Wicked"		Emancipatory Systems Approaches
Ranking	Economic-based decision making			Conflicting/ Coercive	Critical Systems Thinking and Multi-methodology
Routing					

This new version of the GMA maintains the essence of the precedent and gains flexibility in treating different levels of abstraction of the system. It contains the main key parameters and their range values (selected in a mutually exclusive form). They were considered as necessary to address a wide range of alternatives for (re)designing strategies in an organizational system. These parameters and their values, subsequently represented in a matrix, constitute the first two steps in the construction of a Morphological Model [18]. In the next section, the remaining steps are presented.

5 Application of Cross Consistency Assessment and Analysis of the Generated Alternatives

The morphological fields are submitted to judgment through a pair-wise comparison via a cross-impact matrix. The judgment of each pair-wise comparison provides an assessment amongst parameters' values and defines the level of extent of the linkages (in some cases, no linkage) between them. The use of this process, designated as Cross Consistency Assessment (CCA), allows the reduction of the solution space (that contains all the possible configurations of the morphological field) to a subset of internally consistent configurations. This process (also called the "analysis-synthesis process" [18]) goes from an analysis phase – centered on the development of the initial morphological field – to a synthesis phase that provides the representation of a 'solution space', that contains the generated alternative solutions. The application of CCA is useful to detect contradictions and to explore if some combinations may (or not) be appropriate. We led this application by using MA/Carma™, a computational platform that allows the analysis of the configuration produced by the model in a more tractable and faster way. This software belongs to the Swedish Morphological Society, directed by Professor Tom Ritchey. The assessment keys used were: "-" (hyphen), "K" and "X", meaning, respectively: good fit, best fit or optimal pair; possible, could work, but not optimal and, the last one, impossible or very bad idea. A part of this matrix is presented in Fig. 1.

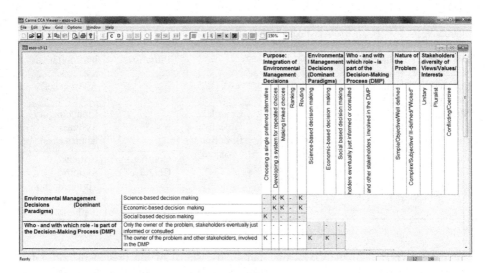

Fig. 1. MA/Carma™: partial screen shot of cross-consistency assessment results.

Despite the fact that these steps are being described in a sequential manner, the process is iterative. The next step is to examine and analyze the alternatives identified automatically by the computational platform. The final iteration resulted in 12 optimal alternatives and in 198 possible alternatives (including the optimal ones). Three possible alternatives are presented in Fig. 2. These alternatives contain different single values that are colorful. The colors code, according to the software in use, attributes the "red" color for "input" variables and the "blue" color for "output" variables. This configuration highlights an interesting characteristic of morphological models: it is possible to choose any of the single values of a parameter as an independent variable or driver. Such characteristic allows defining which are the "input" or "output" drivers, indistinctively and, therefore, exploring the diverse alternatives in an easy manner. It also permits to consider a single variable or a set of variables as "input" simultaneously and then to analyze the behavior of the remaining variables, and its outcomes, leading us to the different alternatives. In our example, we select multiple input variables in order to determine the optimal and feasible alternatives.

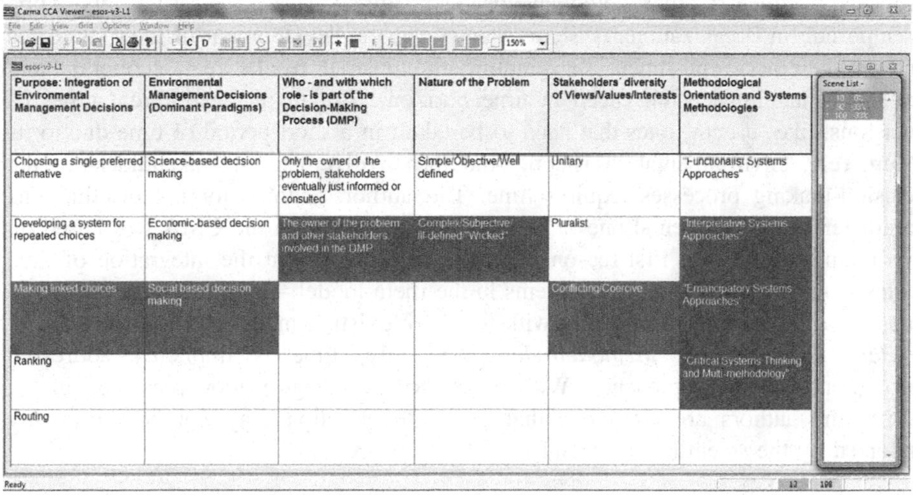

Fig. 2. MA/Carma™: partial screen shot of three possible alternatives in morphological matrix. (Color figure online)

6   Conclusions

In this paper we present a meta-model aiming to support decision-makers that wish to know more about how to use systems models to cope with the integration of environmental concerns into the company strategy. This is made by using a General Morphological Analysis to bridge the gap between OR analysts, decision-makers and stakeholders, making all of them part of the problem structuring and formulation process. In order to deal with a variety of pertinent issues identified in different approaches at the organizational level, we proposed a framework aiming to support decision-making and appraisal of corporate environmental strategies and subsequent management approaches [9, 10]. But there was the need of a methodology for structuring multi-objective problems involving multi-stakeholder environmental decision-making in a participatory context. In this process, it became clear that finding a way of classifying and comparing different types of modeling methods employed in the OR/MS field on the basis of a number of contextual properties would augment the likelihood of success in their application. Like Paucar-Caceres and Espinosa, "we are aware that any categorization carries the risk of placing a methodology in a paradigm that not everybody will accept" [38], but we decided to take this risk.

The presented meta-model is a structured process that allows different paths with diverse possible solutions and provides an audit trail, not only for academic use but also for company decision-makers, contributing to rise its credibility and applicability. "Delivery of models through software or a decision-support system can permit the model to be used by others to make management decisions beyond the timeframe of a scientific research project" [41].

A last note should be done concerning the selection of fundamental parameters and the categorization presented in the GMA. Despite the arguments presented to justify the proposed parameters, others could have been considered. That is the case of *time*. Time is important in organizational systems, namely in the development of their strategies and in their accomplishment for granting the desired results (e.g. economic and environmental impacts unsynced in time horizon). But there are different types of decisions, like 'urgent' ones that need to be taken in a short period of time due to its nature (e.g. environmental disasters). And the involvement of stakeholders in the decision-making processes requires time. The authors' choice for not electing this parameter as a fundamental one to be presented explicitly in the GMA was based on two main reasons. The first reason is related to the fact that the integration of time scales would create additional problems in the meta-model – it would be necessary to integrate a fourth dimension along with the three existing main ones (*Context, Stakeholders and Systemic Methodologies*). Secondly, time is implicitly addressed throughout the OR approaches. With regard to the categorization presented in the GMA, the authors are conscious that this categorization may not be thoroughly endorsed by the scientific community.

Acknowledgments. This work is financed by the ERDF – European Regional Development Fund through the Operational Programme for Competitiveness and Internationalization - COMPETE 2020 Programme within project «POCI-01-0145-FEDER-006961», and by National Funds through the FCT – Fundação para a Ciência e a Tecnologia (Portuguese Foundation for Science and Technology) as part of project UID/EEA/50014/2013.

The authors are also very grateful to the Swedish Morphological Society for supporting the GMA process using MA/Carma™.

References

1. Álvarez, A., Ritchey, T.: Applications of general morphological analysis. Acta Morph. Gen. **4**(1) (2015)
2. Jakeman, A.J., Letcher, R.A., Norton, J.P.: Ten iterative steps in development and evaluation of environmental models. Environ. Modell. Softw. **21**(5), 602–614 (2006)
3. Franco, L.A., Montibeller, G.: Facilitated modelling in operational research. Eur. J. Oper. Res. **205**(3), 489–500 (2010)

4. Ritchey, T.: General morphological analysis: a general method for non-quantified modeling (2013). www.swemorph.com
5. Hämäläinen, R.P., Luoma, J., Saarinen, E.: On the importance of behavioral operational research: the case of understanding and communicating about dynamic systems. Eur. J. Oper. Res. **228**(3), 623–634 (2013)
6. Brocklesby, J.: The what, the why and the how of behavioural operational research—an invitation to potential sceptics. Eur. J. Oper. Res. **249**(3), 796–805 (2016)
7. Hämäläinen, R.P.: Behavioural issues in environmental modelling—the missing perspective. Environ. Model. Softw. **73**, 244–253 (2015)
8. Mingers, J.: Ethics and OR: operationalising discourse ethics. Eur. J. Oper. Res. **210**(1), 114–124 (2011)
9. Teles, M.F., Freire de Sousa, J.: Environmental management and business strategy: structuring the decision-making process in a public transport company. Transp. Res. Proc. **3**, 155–164 (2014)
10. de Fátima Teles, M., de Sousa, J.F.: Integrating environmental policies into business strategy: the problem structuring stage in a framework for decision support. In: Dargam, F., Hernández, J.E., Zaraté, P., Liu, S., Ribeiro, R., Delibašić, B., Papathanasiou, J. (eds.) EWG-DSS 2013. LNBIP, vol. 184, pp. 90–103. Springer, Heidelberg (2014)
11. White, L.: Understanding problem structuring methods interventions. Eur. J. Oper. Res. **199**(3), 823–833 (2009)
12. Lahtinen, T.J., Hämäläinen, R.P.: Path dependence and biases in the even swaps decision analysis method. Eur. J. Oper. Res. **249**(3), 890–898 (2016)
13. Marcus, L., Colding, J.: Toward an integrated theory of spatial morphology and resilient urban systems. Ecol. Soc. **19**(4), 55 (2014)
14. Zwicky, F.: Discovery, Invention, Research Through the Morphological Approach. The Macmillan Company, Collier-Macmillan Canada Ltd., Toronto (1969)
15. Zwicky, F., Wilson, A.: New Methods of Thought and Procedure: Contributions to the Symposium on Methodologies. Springer, Berlin (1967)
16. Ritchey, T.: Problem structuring using computer-aided morphological analysis. J. Oper. Res. Soc. **57**, 792–801 (2006)
17. Müller-Merbach, H.: The use of morphological techniques for OR-approaches to problems. Oper. Res. **75**, 27–139 (1976)
18. Ritchey, T.: Wicked Problems – Social Messes, Decision Support Modelling with Morphological Analysis. Springer, Heidelberg (2011)
19. Warwick, J.: Library operational research: time for a new paradigm? In: Bell, G.A., et al. (eds.) Higher Education Management and Operational Research. Educational Futures, vol. 54, pp. 269–291. Sense Publishers, Dordrecht (2012)
20. Swanich, S.: A critical evaluation of general morphological analysis as a future study methodology for strategic planning. Gordon Institute of Business Science University of Pretoria, Degree Master of Business Administration (2014)
21. Im, K., Cho, H.: A systematic approach for developing a new business model using morphological analysis and integrated fuzzy approach. Expert Syst. Appl. **40**(11), 4463–4477 (2013)
22. Seidenstricker, S., Scheuerle, S., Linder, C.: Business model prototyping–using the morphological analysis to develop new business models. Procedia Soc. Behav. Sci. **148**, 102–109 (2014)
23. Frow, P., Nenonen, S., Payne, A., Storbacka, K.: Managing co-creation design: a strategic approach to innovation. Brit. J. Manage. **26**(3), 463–483 (2015)
24. Storbacka, K., Nenonen, S.: Competitive arena mapping: market innovation using morphological analysis in business markets. J. Bus. Bus. Mark. **19**(3), 183–215 (2012)

25. Heorhiadi, N., Iwaszczuk, N., Vilhutska, R.: Method of morphological analysis of enterprise management organizational structure. ECONTECHMOD Int. Q. J. **2**(4), 17–28 (2013)
26. Ostertagová, E., Kováč, J., Ostertag, O., Malega, P.: Application of morphological analysis in the design of production systems. Procedia Eng. **48**, 507–512 (2012)
27. Jackson, M.C., Keys, P.: Towards a system of systems methodologies. J. Oper. Res. Soc. **35**(6), 473–486 (1984)
28. Jackson, M.C.: Systems Thinking: Creative Holism for Managers. Wiley, New York (2003)
29. Gregory, R., Failing, L., Harstone, M., Long, G., McDaniels, T., Ohlson, D.: Structured Decision Making: a Practical Guide to Environmental Management Choices. Wiley, New York (2012)
30. Ritchey, T.: Four models about decision support modeling. Acta Morph. Gen. **3**(1) (2014)
31. Bryson, J.M.: What to do when stakeholders matter: stakeholder identification and analysis techniques. Publ. Manage. Rev. **6**(1), 21–53 (2004)
32. Daellenbach, H.G.: Hard OR, soft OR, problem structuring methods, critical systems thinking: a primer. In: Conference Twenty Naught One of the Operational Research Society of New Zealand (2001)
33. Paucar-Caceres, A.: Operational research, systems thinking and development of management sciences methodologies in US and UK. Sci. Inq. J. Int. Inst. Gen. Syst. Stud. **9**(1), 3–18 (2008)
34. Checkland, P.B.: Systems Thinking, Systems Practice. Wiley, New York (1981)
35. Schwenk, C.R.: Cognitive simplification processes in strategic decision-making. Strateg. Manage. J. **5**(2), 111–128 (1984)
36. Jackson, M.C.: Systems Approaches to Management. Kluwer Academic Publishers, Dordrecht (2000)
37. Kogetsidis, H.: Systems approaches for organisational analysis. Int. J. Organ. Anal. **19**(4), 276–287 (2011)
38. Paucar-Caceres, A., Espinosa, A.: Management science methodologies in environmental management and sustainability: discourses and applications. J. Oper. Res. Soc. **62**(9), 1601–1620 (2011)
39. Huang, I.B., Keisler, J., Linkov, I.: Multi-criteria decision analysis in environmental sciences: ten years of applications and trends. Sci. Total Environ. **409**(19), 3578–3594 (2011)
40. Macharis, C., Turcksin, L., Lebeau, K.: Multi actor multi criteria analysis (MAMCA) as a tool to support sustainable decisions: state of use. Decis. Support Syst. **54**(1), 610–620 (2012)
41. Kelly, R.A., et al.: Selecting among five common modelling approaches for integrated environmental assessment and management. Environ. Model. Softw. **47**, 159–181 (2013)

Searching for Cost-Optimized Strategies: An Agricultural Application

Christine Largouet[1(✉)], Yulong Zhao[2], and Marie-Odile Cordier[2]

[1] AGROCAMPUS / IRISA, 35042 Rennes Cedex, France
christine.largouet@irisa.fr
[2] Université de Rennes 1/ IRISA, 35000 Rennes Cedex, France

Abstract. We consider a system modeled as a set of interacting components evolving along time according to explicit timing constraints. The decision making problem consists in selecting and organizing actions in order to reach a goal state in a limited time and in an optimal manner, assuming actions have a cost. We propose to reformulate the planning problem in terms of *model-checking* and *controller synthesis* such that the state to reach is expressed using a temporal logic. We have chosen to represent each agent using the formalism of Priced Timed Game Automata (PTGA) and a set of knowledge. PTGA is an extension of Timed Automata that allows the representation of cost on actions and the definition of a goal (to reach or to avoid). This paper describes two algorithms designed to answer the planning problem on a network of agents and proposes practical implementation using model-checking tools that shows promising results on an agricultural application: a grassland based dairy production system.

Keywords: Decision support system · Temporal planning · Optimized planning · Timed automata · Model-checking

1 Introduction

When managing agro-ecosystems, one of the major decision-aid problems is to find the best management strategy to ensure the health, resilience and diversity of the ecosystem. In this paper we address the problem of finding a strategy for a class of systems that encompasses the characteristics of agro-ecosystems. We investigate the category of non-deterministic systems, defined as a group of interacting agents, each one having its proper dynamics submitted to explicit timing constraints. To improve agro-ecosystem management the model has to include the interactions between environmental, human and ecological subsystems. Time is another fundamental characteristic to express the dynamical features of these systems which are traditionally modeled as differential equations in quantitative approaches. Additionally to interaction and time, we are interested in two novel issues: the representation of unexpected events such as climatic events (hurricane, flooding, heat wave, etc.) and the expression of cost on actions to handle

© Springer International Publishing Switzerland 2016
S. Liu et al. (Eds.): ICDSST 2016, LNBIP 250, pp. 31–43, 2016.
DOI: 10.1007/978-3-319-32877-5_3

the environmental impact produced by human activities. Searching for a strategy in these systems consists of choosing and organizing the actions through time in order to achieve an optimal goal, assuming that the actions of the agents have a cost.

To cope with the complexity of systems such as large ecosystems, we propose to use symbolic methods applied with success in model-checking. The paradigm of *planning with model-checking* first proposed by [1,2] is relatively recent and generated numerous papers on a variety of planning domains. These works have emphasized that this promising approach can tackle the problem of generating plans on nondeterministic models for extended goals. Some other work on timeline-based planning has introduced explicit time on models for space applications [3,4]. At the same time, in the multi-agent systems community, Alternating-Time Temporal Logic (ATL) has been used for reasoning about agents and their strategies [5] and has been made available in model-checkers such as MCMAS [6]. Recent work has focused on the verification against epistemic logic, or logic for knowledge [7,8] but one of the key issues remains the state-space explosion problem.

In this paper, we propose to reformulate the planning problems in terms of *model-checking* and *controller synthesis* associated to a temporal logic to express the goal to reach or to avoid. Model-checking and controller synthesis have been widely studied for discrete event systems and more particularly for timed automata [9]. In our approach we propose to represent the system as a network of interacting agents, each agent being described as an extended timed automaton and a set of knowledge. The formalism we chose is Priced Timed Game Automata (PTGA), an extension of timed automata that allows to express timing constraints on states and transitions, costs on actions and the definition of goals [10]. This representation is relevant for multi-agent systems since the interaction can be performed through communicating and synchronizing actions.

On a network of PTGA, we propose an approach to formulate the planning strategies but also to compute the cost of each strategy. The requirement expressed in a temporal logic asks the following question: "What is the best strategy to guide the system to a specific goal at a specific time?". In our case, "best" means that a criterion should be minimized and this criterion is the strategy cost. We propose two strategy search algorithms relying on efficient and recognized tools for the model-checking and the synthesis. The first algorithm focuses on searching for the best strategy on a multi-agent system, whereas the second combines controller synthesis and machine learning in order to generate a meta-strategy for a class of similar multi-agent systems.

This approach is applied on grassland management where reasonable practices in farming systems are essential for sustainable agriculture. Land use changes associated with intensive practices such as abusive use of fertilizer or increased number of cutting and grazing activities could have severe impacts on environmental systems. Most of the models, which are designed for agro-ecosystem management, focus on the grassland simulation but can not calculate explicit grassland management strategies. We propose a tool named PaturMata implementing our method and dedicated to the exploration of the cost-optimal grassland management strategies in a dairy production system.

The paper is structured as follows. Section 2 introduces the formalism of PTGA. Section 3 presents some of the background in model-checking and controller synthesis. Section 4 describes the network of agents and describes the two algorithms that lead to the strategies. Section 5 briefly describes the application of this approach on ecological systems: grass-based dairy farming. Section 6 concludes and outlines directions for future research.

2   Priced Timed Game Automata (PTGA)

Clock Constraints. Timed Automata [9] are automata enriched with a set of variables called clocks. Let $\mathcal{X}$ be a set of *clocks*. A *(clock) valuation* v for a set $\mathcal{X}$ assigns a real value to each clock. The set of *clock constraints* over $\mathcal{X}$, denoted $\Phi(\mathcal{X})$ is defined by the grammar : $\varphi ::= x \leq c \mid c \leq x \mid x < c \mid c < x \mid \varphi_1 \wedge \varphi_2$ where φ, φ_1 and φ_2 belong to $\Phi(\mathcal{X})$; $x \in \mathcal{X}$ is a clock and $c \in \mathbb{R}^+$ a constant. Clock constraints are evaluated over clock valuations. A constraint φ can be viewed as the set of valuations that satisfy it, hence, we say that v satisfies φ, denoted $v \vDash \varphi$, if $v \in \varphi$.

Timed Automata and Extended Formalisms. In timed automata [9], the vertices of the graph are called *locations*, the clock constraint associated to a location is called an *invariant*, whereas the one associated to an edge is called a *guard*. An edge is decorated with an event label and allows the resetting of clocks. In a location, a transition can be triggered if the guard of the outgoing edge is satisfied. The triggering of a transition is instantaneous and takes no time. Priced Timed Automata (PTA) [11] extend the formalism of timed automata by adding cost to the behavior of timed systems. In PTA, locations and edges are annotated by cost. The cost label on a location represents the price per time unit while staying in this location whereas the cost label on an edge expresses the cost when the transition is triggered. Timed Game Automaton (TGA) [12] is an extension of timed automata where actions on edges are partitioned into controllable and uncontrollable actions. Based on these principles, Priced Timed Game Automata (PTGA) is an extension of both PTA and TGA [10].

2.1   PTGA Definition

A Priced Timed Game Automaton $\mathcal{A}$ is a tuple $< \mathcal{S}, \mathcal{X}, Act, \mathcal{E}, \mathcal{I}, \mathcal{P} >$ where:

- $\mathcal{S}$ is a finite set of locations and $s_o \in S$ is the initial location.
- $\mathcal{X}$ is a finite set of clocks.
- $Act = Act\_c \cup Act\_u$ is a finite set of actions divided into Act_c, the *controllable* actions (played by the controller), and Act_u, the *uncontrollable* actions (played by the environment).

- $\mathcal{E} \subseteq \mathcal{S} \times \mathcal{A}ct \times \varPhi(\mathcal{X}) \times 2^{\mathcal{X}} \times \mathcal{S}$ is a finite set of edges. Each edge e is a tuple $(s, l, \varphi, \delta, s')$ such that e connects the location $s \in \mathcal{S}$ to the location $s' \in \mathcal{S}$ on the event label $l \in \mathcal{A}ct$. The enabling condition (called the *guard*) is captured in $\varphi \in \varPhi(\mathcal{X})$. $\delta \subseteq \mathcal{X}$ gives the set of clocks to be reset when the transition is triggered.
- $\mathcal{I} : \mathcal{S} \rightarrow \varPhi(\mathcal{X})$ maps each location s with a clock constraint called an *invariant*.
- $\mathcal{P} : \mathcal{S} \cup \mathcal{E} \rightarrow \mathbb{N}$ assigns cost rates and costs to locations and edges, respectively.

Figure 1 presents a PTGA having one clock x that is used to define invariants and guards, colored in green in the figure. The goal to reach is the state $s4$. Cost-rates, $cost'$, expresses the cost per time unit while staying in the referred location whereas the cost of triggering a transition is defined by $cost$. Two edges are associated to uncontrollable actions: $a3$ and $a5$ (the edges are shown with a dotted line).

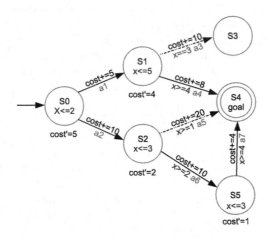

Fig. 1. A PTGA having one clock x, two uncontrollable actions ($a3$, $a5$), costs on locations and edges and a goal $s4$.

The semantics of a PTGA is a *priced transition system* whose states $q \in \mathcal{Q}$ are the pairs $q = (s, v)$ such that $s \in \mathcal{S}$, is a location, and $v \in \mathbb{Q}$ is a clock valuation where v satisfies the invariant $\mathcal{I}(s)$ of the location. We write $q \xrightarrow{l}_c q'$ the transition between two states q and q' labeled by the action l while c is the cost of this transition.

From a current state, a PTGA can evolve into a destination location through one of the outgoing edges or remains in its current location while time passes. Consequently, two kinds of transitions are distinguished: discrete and delay transitions. A *discrete transition* is enabled if the timing information of the edge is satisfied (the guard). The cost value does not impact the triggering of the edge. In a discrete transition, the cost of the transition is the cost of the triggered edge. A *delay transition* can be performed if the invariant of the active transition is satisfied. The cost of a delay transition is computed as the product of the duration and the cost rate of the current location.

A *finite run* ϱ of a PTGA called A is a finite sequence of transitions, starting from the initial state: $q_0 \rightarrow^1 q_1 \rightarrow^2 q_2 \rightarrow^3 \cdots \rightarrow^n q_n$. The cost of ϱ, denoted $cost(\varrho)$ is the sum of all the costs along the run.

$$cost(\varrho) = \sum_{1 \leq i \leq n} \begin{cases} \delta \cdot \mathcal{P}(s) & \text{if } \rightarrow^i \text{ is a delay transition} \\ \mathcal{P}(e) & \text{if } \rightarrow^i \text{ is a discrete transition} \end{cases}$$

The minimum cost of reaching a location s is the minimum cost of all the finite runs from s_0 to s.

Let us consider the PTGA $\mathcal{A}$ given Fig. 1, a possible run in $\mathcal{A}$ is:

$$\varrho : (s_0, 0) \xrightarrow{\delta(2)} (s_0, 2) \xrightarrow{a1} (s_1, 2) \xrightarrow{\delta(1)} (s_1, 3) \xrightarrow{a4} (s_4, 3)$$

The cost of ϱ is $cost(\varrho) = 5 \times 2 + 5 + 8 \times 1 + 15 = 38$.

2.2   Best-Cost Strategy Problem for PTGA

Given a PTGA and a goal state, the best-cost strategy problem is to find a run, with the optimal cost, starting from the inital state and leading to the goal state. For the example of Fig. 1, a key issue is for instance: "Does the controller activate the action $a1$ or $a2$ from the initial state $S0$ to reach in an optimal way the goal $S4$?". If the preferred run takes the edge labelled with $a2$, another question could be: "Is it possible to win whatever the behavior of the environment?".

3   Model-Checking and Controller Synthesis

To face the large combinatorial size of the state-space, symbolic efficient data structures provide a very compact encoding for large sets of states [13]. When a system is described as a network of timed automata (extended or not), it can be treated using two successful techniques: *model-checking* [14] and *controller synthesis* [15]. These both techniques require the expression of a property to be satisfied and expressed in a temporal logic. The most popular is TCTL [16], a convenient formalism to specify properties over timed automata.

3.1   Timed Computation Tree Logic (TCTL)

The grammar of TCTL is the following:

$$f ::= p \mid x \in I \mid \neg p \mid p_1 \vee p_2 \mid \exists \Diamond_I p \mid \forall \Diamond_I p \mid \exists \Box_I p \mid \forall \Box_I p$$

where p is a property, $x \in \mathcal{X}$ is a clock and I is a time interval. I is an interval with integer bounds of the form $[n, n']$ with $n, n' \in \mathbb{N}$. The diamond operator $\Diamond p$ expresses that a path (i.e. a sequence of states) leads to a state satisfying the property p. The box operator $\Box p$ means that all the states along a path satisfy the property p. These modal operators can be combined with the universal quantifiers $\exists$ or $\forall$ over the paths. The formula $\exists\Diamond_{[0,3]}\, p$ expresses that there is at least one path leading to a state satisfying p within 3 units of time.

3.2   Model-Checking

Model-checking is performed using efficient tools called *model-checkers* dedicated to answer whether or not a property is satisfied by the system. The property is expressed using specification languages such as temporal logics. The problem of model-checking can be expressed as follows: given a system model M and a property φ to be checked, does the model M satisfy φ?

3.3   Controller Synthesis

Controller synthesis is the problem of finding a way to control the system so that the behavior of the system satisfies a given property. The objective is then to *synthesize* a controller. This controller coupled with the system has to respect the given specification. On a PTGA, we denote by Σ the set of possible actions that could be proposed by the controller so that $\Sigma = Act\_c \cup \lambda$ with Act_c the set of controllable actions and λ the action of letting time pass. A strategy is winning if, when following these rules, the controller always wins whatever the environment does (by the way of non-controllable actions).

Definition 1 (Decision Rule). *A decision rule gives an action to be performed on a specific system state at a particular time. A decision rule is a tuple (s, φ, σ) such that $s \in \mathcal{S}$ is a location of the PTGA, $\varphi \in \Phi(\mathcal{X})$ is a clock constraint and $\sigma \in \Sigma$ is a controllable action.*

A controller C_f on a PTGA called $\mathcal{A}$ is a system so that coupled with $\mathcal{A}$ controls the behavior of $\mathcal{A}$ according to a strategy f. We denote by $C_f \,\|\, \mathcal{A}$, the PTGA $\mathcal{A}$ controlled by C_f. Controller synthesis distinguishes two kinds of control objectives depending on the property Φ to satisfy: *reachability* and *safety* that can be expressed as following:

- *Reachability:* Given a PTGA $\mathcal{A}$ and a property φ to reach, the controller synthesis is to find a strategy f such that $(C_f \,\|\, \mathcal{A}) \models \forall\Diamond\varphi$.
- *Safety:* Given a PTGA $\mathcal{A}$ and a property φ to avoid, the controller synthesis is to find a strategy f such that $(C_f \,\|\, \mathcal{A}) \models \forall\Box\neg\varphi$.

Several strategies can potentially fulfill the property Φ to satisfy. We call $C^* = \{C_f \mid (C_f \,\|\, \mathcal{A}) \models \Phi\}$ the set of possible controllers. Two of them are of interest:

- *Complete controller:* The complete controller $C_{fmax} \in C^*$ corresponds to the *complete strategy* containing all the possible decision rules. The controller C_{fmax} is related to the complete strategy f_{max} such that $\forall C_f \in C*, f \subseteq f_{max}$. In a complete strategy, more than one action can be potentially eligible for each location. The controller C_{fmax} needs to choose one action among all the possible ones.

– *Minimal controller:* The minimal controller C_{fmin} corresponds to the *minimal strategy* such that: $\neg \exists C_f \in C^* \mid f \subset f_{min}$. A minimal strategy is a minimal set of decision rules. In that case, there is only one decision rule for a location. A minimal strategy is not necessary unique.

The cost of a strategy f from (s, v) is defined by:
$cost(f, (s, v)) = sup \{cost(\varrho \mid \varrho \in Outcome(f, (s, v))\}$. The output of a strategy f, denoted $Outcome((s, v), f)$, from a state (s, v) is a subset of all the runs starting from (s, v) and satisfying f (see complete definition in [10]).

3.4 Tools

UPPAAL [17] is a collection of tools dedicated to the analysis of timed automata and its extended formalisms. In all tools, properties are expressed using the logic TCTL. UPPAAL TIGA [18] performs controller synthesis on timed game automata (TGA) with respect to reachability and safety properties. UPPAAL TIGA can provide a complete controller or one of the minimal controllers, randomly chosen among all the possible minimal controllers. When dealing with priced timed automata (PTA), UPPAAL CORA [19] is a model-checking tool that can be used to explore all runs that answer a reachability property in order to retrieve the optimal controller.

4 Strategy Search Methods

We propose two methods to compute the planning strategies. The first one called *Best Strategy Search* is looking for the optimal strategy on a model, whereas the second, *Meta-Strategy Search*, provides a meta-strategy for a class of models.

4.1 Network of Agents

The *strategy search* methods are applied on multi-agent systems represented by a set of interacting agents, each agent being described by a PTGA and complementary knowledge.

Knowledge Descriptors: Agents are enriched with a set of variables called *knowledge descriptors*. The knowledge descriptors define global information on the agent and its behavior. Let KD be a set of knowledge descriptors. Each agent A_i defines a valuation over $KD_i \subseteq KD$ by assigning an integer value to each knowledge descriptor.

Model: A model consists of a network of n interacting agents A_i. Each agent is composed of two parts: a PTGA, $PTGA_i$ and a set of knowledge descriptors KD_i. A model $\mathcal{M}$ is defined as follows: $\mathcal{M} = \{A_i :< PTGA_i, KD_i >\}$ such that:

- $PTGA_i$ is a PTGA defining one agent A_i.
- KD_i define the set of valuations for each knowledge descriptor such that $KD_i \subseteq KD$.

The interaction between the agents is realized by the PTGA according to the CSS parallel composition operator [20] that allows interleaving of actions and hand-shake synchronization (on specified actions through communication channels).

4.2   Best Strategy Search Algorithm

Our strategy search algorithm uses the efficient UPPAAL tools that only rely, for computational reasons, on timed game automata (TGA) and priced timed automata (PTA). Given this constraint, we make an assumption in the design of the PTGA: the non-determinism should be only restricted to controllable actions. Given a PTGA as a model, PTA and TGA can be easily derived according to the following definitions.

Definition 2 (PTA derived from a PTGA). *A Priced Timed Automaton (PTA) is derived from a PTGA if all the actions, controllable and uncontrollable, that are labelling the edges are replaced by classical event labels $e \in \mathcal{A}ct$, with $\mathcal{A}ct$ now defined as a set of classical event labels. Events usually label the edges of a timed automaton when any notion of controllability (or uncontrollability) is required.*

Definition 3 (TGA derived from a PTGA). *A Timed Game Automaton (TGA) is derived from a PTGA by removing all the cost, both on locations and edges.*
Nota: We denote by $\oplus$ the operation that add discrete variables, as authorized in UPPAAL, to a classical TGA. A variable called varCost can be added to a derived TGA, A, such that $A \oplus varCost$ enables the definition of the costs on A from the original PTGA.

The Algorithm 1 *BestStratSearch* computes on a model $\mathcal{M}$ the optimal strategy to reach a specific goal expressed by the system state g. This strategy (called *Strategy* in the algorithm) corresponds to the complete controller. The algorithm is only based on the PTGA part of each agent A_i of the model $\mathcal{M}$. The principle of the Algorithm 1 follows two steps: (1) the search for the optimal cost and (2) the computation of the strategy corresponding to this cost. In a first part, the PTGA of each agent is derived in a PTA and the synchronized product is computed. The model-checker UPPAAL CORA is called with in input, the synchronized product of the model and a TCTL formula meaning "Is there is path leading to the state g?". If this path exists, the model-checker returns the optimal cost called *OptCost*. In a second step, the algorithm derives the model in a set of TGA extended with a variable of cost, *varCost*, and computes the synchronized product of TGA. UPPAAL TIGA searches for the strategy corresponding to the previously found optimal cost. The strategy provided by UPPAAL TIGA as a result of this algorithm is the complete strategy associated to the complete controller, a strategy defined as a set of decision rules: $\{(s, \varphi, \sigma)\}$.

Algorithm 1. BestStratSearch

Require:
- $\mathcal{M} = \{A_i = <PTGA_i, KD_i>\}$, /* $i \in [1, n]$ with n the number of agents */
- $g \leftarrow state\_to\_reach$

1- Search for the best cost
for all $PTGA_i$ **do**
  $PTA_i \leftarrow DerivedPTA(PTGA_i)$
end for
$Prod1 \leftarrow SynchroProduct(\{PTA_i\})$
$Query1 \leftarrow "\exists \lozenge \, g"$
$OptCost \leftarrow UppaalCora(Prod1, Query1)$

2- Search for the controller related to the best cost
for all $PTGA_i$ **do**
  $TGA_i \leftarrow DerivedTGA(PTGA_i) \oplus varCost$
end for
$Prod2 \leftarrow SynchroProduct(\{TGA_i\})$
$\varphi \leftarrow varCost = OptCost$
$Query2 \leftarrow "\exists \lozenge g \wedge \varphi"$ /* Is there a run leading to g such that $varCost$ equals the optimal cost $OptCost$ previously computed */
$Strategy \leftarrow UppaalTiga(Prod2, Query2)$

4.3   Meta-Strategy Search Algorithm

Similar Models: Two models $\mathcal{M}_1$ and $\mathcal{M}_2$ are *similar* if they only differ by the values of their knowledge descriptors KD_i of each agent A_i of the model. It means that $\mathcal{M}_1$ and $\mathcal{M}_2$ are composed of the same $PTGA_i$ and their KD_i define the same variables, only their values differ.

The previous method provides the best and complete strategy for one model $\mathcal{M}$. However, the stakeholder is sometimes more interested by having general rules that work on similar models. This is the case in agro-ecosystem management where systems can not be defined very precisely because parameters are unknown or data unavailable. In such situations, working on a class of similar models is less restrictive and allows the computation of more general decision rules. These generalized rules can be viewed as meta-strategies and are extracted from a set of similar models. However, the generalization of the rules may lead to a minor loss of the optimality. It means that the application of the generalized rules do not always give the optimal cost. For agro-ecosystem management, stakeholders are interested by more interpretable rules than too specific ones that are difficult to apply.

Given a goal to achieve, Algorithm 2, *MetaStratSearch*, explains how to compute the meta-strategy on a set of similar models $\mathcal{M}_j$ with $j \le m$ and m the number of models. Each model is composed of a set of interacting agents A_i with $i \le n$ and n the number of agents. We propose to compute the optimal strategy for each model $\mathcal{M}_j$ using the previous algorithm *BestStratSearch* for the goal to reach g. Each optimal strategy $Strat_j$ provided by this algorithm is associated to the global set of knowledge descriptor values MKD_j such that $MKD_j = \cup_{i \in [1,n]} KD_i$ with n the number of agents of each model $\mathcal{M}_j$. Each couple $(Strat_j, MKD_j)$ is added in a strategy base called $StrategyBase$. Hence, this base of strategies is exploited by a machine learning algorithm to provide the meta-strategy. We chose to apply a classical rule learner algorithm, Rip-per (Repeated Incremental Pruning to Produce Error Reduction), designed to generate rule sets for datasets with many features [21].

Algorithm 2. MetaStratSearch

Require:
  - $\{\mathcal{M}_j\}$ /* Similar models */
  - $g \leftarrow state\_to\_reach$
  for all $\mathcal{M}_j$ **do**
    $Strat_j \leftarrow BestStratSearch(\mathcal{M}_j, g)$
    $MKD_j \leftarrow \cup_{i \in [1,n]} KD_i$, /* n the number of agents of $\mathcal{M}_j$ */
    $StrategyBase \xleftarrow{+} (Strat_j, MKD_j)$
  end for
  $MetaStrategy \leftarrow Learner(StrategyBase)$

The meta-strategy provided by this algorithm is expressed as a set of generic rules, valid for the models.

5   Agricultural Application

We applied our approach on a grassland based dairy production system using a prototype software named PaturMata. PaturMata comes from the combination of the two words *paturage* (which is a french word meaning pasture) and *automata*. In this context the main issues are to maintain the dairy production at a desired level while limiting the nitrogen fertilization used to increase the grass growth which is known for its environmental damages. PaturMata[1] is a decision support system that models the grass growth under different climate condition, grass consumption by the herd and some agricultural activities like grass cutting and soil fertilization. The model in PaturMata is composed by several agents which are organized in a hierarchy according to their functions (cf. Fig. 2):

- Grassland layer: This is the biological model which simulates the grass growth and consumption in each paddock.
- Execution layer: This is the activity model which represents all the agricultural activities including herds' movement, grass cutting and soil fertilization. Each activity is defined by a PTGA. A grazing PTGA represents one herd and

---

[1] PaturMata can be freely downloaded from the website: http://people.irisa.fr/ Christine.Largouet/paturmata.html.

simulate its movement from one paddock to another. A grass cutting PTGA and a fertilization PTGA simulate the cutting and the fertilization activity on one specific paddock.

– Controller layer: The different models of this layer simulate the strategies management. One PTGA of the controller layer is associated to a PTGA of the execution layer. To model human decisions, these automata are activated once a day.

– Time layer: One PTGA is in charge of the scheduling.

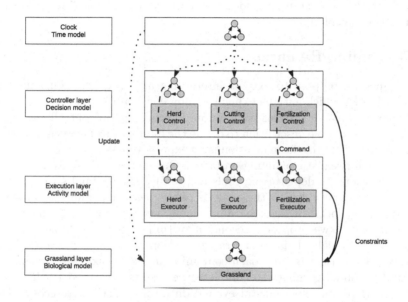

Fig. 2. PaturMata architecture in 4 layers composed of PTGA

We carried out one experiment on 100 similar paddock configurations. 50 of them are used to generate fertilization strategy and the other 50 are used to test the strategy. The knowledge descriptor associated to each configuration are: surface of paddocks, distance between paddock and the milking building. We present here some of the decision rules obtained from the *MetaStratSearch* algorithm.

– Fertilize a paddock
  – if the date is between the 15th Mars and the 1st April, and
  – the distance between the paddock and the milking building is between 400 and 800 m, and
  – the surface of the paddock is larger than 2.5 ha.
– Fertilize a paddock
  – if the date is between the 15th Mars and the 1st April, and
  – the surface of the paddock is between 2.0 and 2.2 ha.

- Fertilize a paddock
    - if the date is between the 15th Mars and the 1st April, and
    - the surface of the paddock is between 1.5 and 1.8 ha, and
    - the distance between the paddock and the milking building is less than 400 m.

To test the performance, we applied the obtained strategy on the 50 paddock configurations of the test set and noted the cost at the end of simulation. We then launch *BestStratSearch* algorithm on each configuration in order to calculate the optimal cost. A simple comparison showed that the cost obtained by applying the meta-strategy on the model doesn't exceed 20 % of the optimal cost on 39 configurations out of 50.

6   Concluding Remarks

In this paper we propose to express interacting agents in the convenient formalism of PTGA which gathers explicit timing constraints and cost on actions. PTGA allows to model systems having non-determinism on controllable actions and offers a manageable description to define the interactions between the agents. If the main benefit of this formalization is its expressiveness, these models are not easy to tackle for realistic planning problems. Our method proposes using recognized and efficient model-checking tools to produce the optimal strategy. However, the strategy provided by the analysis of one particular multi-agent system is sometimes too specific when regarding the agro-ecosystem management problem. Consequently, we propose a second algorithm to generate a meta-strategy from a class of models. This meta-strategy is more easily interpretable. To our knowledge, this is the first approach combining model-checking, controller synthesis and machine learning. To complete the results presented in this paper, we can report on an experimental evaluation on a realistic agro-ecosystem: a grassland based dairy production system.

One limitation of this work is that the expression of the non-determinism is only effective for controllable actions. This is a consequence of the derivation of PTGA to PTA where all the transitions are considered as controllable. A first perspective to consider for future work would be to study how to avoid this limitation in order to exploit the full potential of the expressiveness of this formalism. To improve the algorithm performance, an idea would be to consider the reduction of the memory used by the search algorithm. This could be realized by using, in the PTGA, clocks dealing with integer values instead of real values. This granularity of time is sufficient for the management of agro-ecosystems. Finally, a further perspective would be to validate PaturMata through long-term real-life agricultural practices.

References

1. Giunchiglia, F., Traverso, P.: Planning as model checking. In: Biundo, S., Fox, M. (eds.) Recent Advances in AI Planning. LNCS, vol. 1809, pp. 1–20. Springer, Heidelberg (1999)

2. Cimatti, A., Pistore, M., Roveri, M., Traverso, P.: Weak, strong, and strong cyclic planning via symbolic model checking. Artif. Intell. **147**(1–2), 35–84 (2003)
3. Cesta, A., Finzi, A., Fratini, S., Orlandini, A., Tronci, E.: Analyzing flexible timeline-based plans. In: ECAI-2010, pp. 471–476 (2010)
4. Orlandini, A., Finzi, A., Cesta, A., Fratini, S.: Tga-based controllers for flexible plan execution. KI **2011**, 233–245 (2011)
5. Lomuscio, A., Raimondi, F.: Model checking knowledge, strategies, and games in multi-agent systems. In: AAMAS-2006, pp. 161–168 (2006)
6. Lomuscio, A., Qu, H., Raimondi, F.: MCMAS: a model checker for the verification of multi-agent systems. In: Bouajjani, A., Maler, O. (eds.) CAV 2009. LNCS, vol. 5643, pp. 682–688. Springer, Heidelberg (2009)
7. Lomuscio, A., Michaliszyn, J.: Decidability of model checking multi-agent systems against a class of EHS specifications. In: ECAI-2014, pp. 543–548 (2014)
8. Huang, X., Van der Meyden, R.: Symbolic model checking epistemic strategy logic. In: AAAI-2014, Québec, Canada (2014)
9. Alur, R., Dill, D.: A theory of timed automata. Theoret. Comput. Sci. **126**, 183–235 (1994)
10. Bouyer, P., Cassez, F., Fleury, E., Larsen, K.G.: Optimal strategies in priced timed game automata. In: Lodaya, K., Mahajan, M. (eds.) FSTTCS 2004. LNCS, vol. 3328, pp. 148–160. Springer, Heidelberg (2004)
11. Alur, R., La Torre, S., Pappas, G.J.: Optimal paths in weighted timed automata. In: Di Benedetto, M.D., Sangiovanni-Vincentelli, A.L. (eds.) HSCC 2001. LNCS, vol. 2034, pp. 49–62. Springer, Heidelberg (2001)
12. Asarin, E., Maler, O., Pnueli, A.: Symbolic controller synthesis for discrete and timed systems. In: Antsaklis, P., Kohn, W., Nerode, A., Sastry, S. (eds.) Hybrid Systems II. LNCS, vol. 999, pp. 1–20. Springer, Heidelberg (1995)
13. Henzinger, T., Nicollin, X., Sifakis, J., Yovine, S.: Symbolic model checking for real-time systems. Inf. Comput. **111**(2), 193–244 (1994)
14. Clarke, E., Grumberg, O., Peled, D.: Model-Checking. MIT Press, USA (2002)
15. Ramadge, P., Wonham, W.: The control of discrete event systems. IEEE **77**, 81–98 (1994)
16. Alur, R., Courcoubetis, C., Dill, D.L.: Model-checking in dense real-time. Inf. Comput. **104**(1), 2–34 (1993)
17. Larsen, K., Pettersson, P., Yi, W.: Uppaal in a nutshell. Int. J. Softw. Tools Technol. Transfer **1**, 134–152 (1997)
18. Behrmann, G., Cougnard, A., David, A., Fleury, E., Larsen, K.G., Lime, D.: UPPAAL-Tiga: Time for Playing Games! In: Damm, W., Hermanns, H. (eds.) CAV 2007. LNCS, vol. 4590, pp. 121–125. Springer, Heidelberg (2007)
19. Behrmann, G., Larsen, K.G., Rasmussen, J.I.: Priced timed automata: algorithms and applications. In: de Boer, F.S., Bonsangue, M.M., Graf, S., de Roever, W.-P. (eds.) FMCO 2004. LNCS, vol. 3657, pp. 162–182. Springer, Heidelberg (2005)
20. Milner, R.: Communication and Concurrency. Prentice Hall, New York (1989)
21. Cohen, W.: Fast effective rule induction. In: Proceedings of the Twelfth International Conference on Machine Learning, pp. 115–123. Morgan Kaufmann (1995)

Enhancing Antenatal Clinics Decision-Making Through the Modelling and Simulation of Patients Flow by Using a System Dynamics Approach. A Case for a British Northwest Hospital

Jorge E. Hernandez[1,4(✉)], Ted Adams[2], Hossam Ismail[3], and Hui Yao[1]

[1] University of Liverpool Management School, Liverpool, UK
J.E.Hernandez@Liverpool.ac.uk
[2] Liverpool Women's Hospital, Liverpool, UK
tedadams@doctors.org.uk
[3] International Business School Suzhou (IBSS), Xi'an Jiaotong-Liverpool University (XJTLU), Suzhou, China
[4] Universidad de La Frontera, Temuco, Chile

Abstract. In the past 60 years, the maternal mortality rate in the United Kingdom has dropped considerably. However, the number of high-risk pregnancies including those complicated by pre-existing maternal health problems e.g. diabetes or lifestyle illnesses e.g. obesity has resulted in an increased demand on obstetric outpatient management of pregnancy at British National Health Service hospitals. In addition, patients also expect better access and convenient appointments in the antenatal clinic. Despite on-going work in these areas, long delays in clinic waiting rooms continue to be a great source of frustration for patients and staff. These delays have a considerable social cost to the economy and a financial cost to the health economy. Therefore, this paper considers a realistic study for supporting decision-makers in antenatal clinics in British northwest hospitals by using a system dynamics approach through causal-loop diagrams. The focus is to enhance the performance of the clinic, by understanding the flow of patients though a hospital clinic thereby aiming to reduce waiting times for patients.

Keywords: Healthcare · Decision-making · Operations management · System dynamics simulation

1 Introduction

In the early twentieth century, antenatal care programs have been widely adopted throughout many countries. These countries have had a greater coverage rates in skilled birth attendants and in the postnatal period [6]. During the last decade, the World Health Organization have championed antenatal care coverage (ANC) which is specifically antenatal care and is used in many countries. It suggests that practitioners offer pregnant women at least four visits which are an opportunity to provide essential tests, for example

© Springer International Publishing Switzerland 2016
S. Liu et al. (Eds.): ICDSST 2016, LNBIP 250, pp. 44–55, 2016.
DOI: 10.1007/978-3-319-32877-5_4

to test for HIV and to talk about health improvement messages such as diet and smoking advice [7]. Currently, this is considered a successful program across the globe with 71 % of women worldwide and over 95 % of women in industrialized countries having access to ANC. In Africa, 69 % of women have at least one ANC visit [7]. In the United Kingdom (UK), NHS (National Health Service) organisations usually follow the NICE guideline [12] for Antenatal Care. The guideline sets out the care that pregnancy women should have. The care includes screening measures, preventative care and ongoing management of other medical problems that impinge or are altered by pregnancy. Therefore, when women are pregnant, they make contact with a local midwife who arrange a programme of antenatal care. Women in their first pregnancy will usually have at least 10 appointments. The first appointment, called the booking appointments, is an opportunity for a local midwife, to find out about this and previous pregnancies. They would normally find out about the current health of women and babies, undertake screening tests (e.g. mental health) and provide useful information around healthy. It is also an opportunity for women and their partners to ask questions and to find out about antenatal classes [8]. Women who have medical conditions are often classed as "high risk". If this is the case, their antenatal care will usually be shared between an obstetrician and a midwife. Many professionals make up the antenatal care team, each providing expertise in their own area. These professionals include obstetricians, anaesthetists, neonatologists and midwives who together provide a comprehensive service on the basis of patients' requirements and problems, thus ensuring the lowest risks for women and babies [8]. In this paper, we analysed a large antenatal clinic in the NHS which is typical of Consultant led antenatal clinics in the UK.

The research covered in this paper aims to provide an understanding of antenatal care and its important role in the world health, including the managerial and control of the operational processes, although, and even more important, to use system dynamics to realise what are the main delay factors that affects the waiting times in these antenatal clinics. The structure of this paper is as follows. In the first section we explore the background of antenatal health care with a view on system dynamics modelling. In the second section the system dynamic is addressed, which presents the causal-loop diagram and how this supports the decision-making process in the Hospital Antenatal Clinics (HANC). Finally, and based on real data coming from northwest region from UK, specific simulation scenarios and it's outcomes is presented. Hence, and based on the experience from NHS participants, this paper is generalizable to the way on how many HANC's are set up across the UK, using antenatal care as an exemplar.

2 Background

System dynamics (SD), is a computer-based simulation approach, which has a long history of utilization in various domains, such as management, in the manufacturing and energy industries, for example [4]. Related to this, [1] indicates that SD seems to stand in a top-down platform with the strategic foresight compared to Discrete-event-simulation. In fact, [11] selected the simulation method to analyse the impact of elderly patients on the further high budget in NHS. In this context, Healthcare organizations, could use SD in capacity planning, patient behavior management and resource allocation. With a focus on systematic structure in analysing healthcare, SD could relate all relevant factors and healthcare interactions to each other instead of analysing these elements individually [1, 2], which suggests the lack of a comprehensive and systematic perspective [3]. In this context, the system dynamics has been widely used in various industries including healthcare to help organizational managers address dynamic problems in complex environments. In the case of healthcare, system dynamics can be used to analyse the interdependence between parts of the system, thereby generating feedback from models [10]. The literature around the use of system dynamics in healthcare can be divided into two parts (1) analysis of specific diseases and (2) operational management including application of how operational policies affect an organisation. For example, the demands placed on health services because of population growth o an ageing population and limitation resources Place additional pressures on the healthcare system as a whole. Issues such as long waiting time for treatment can result in unsatisfactory service quality and many complaints from patients. Researchers have used a system dynamics model to analyse the operation of Emergency Departments. [9] and also used SD to simulate the interaction between emergency care systems and the social care system, [5] SD modelling has been used to analyse the effects of hospital bed reduction on the operation. Through these models, researchers tested whether the assumption of reducing inpatient beds would improve delays in discharge. But, more important, SD was also used to test the relationship between patients flow and waiting times for hospital beds. Moreover, [1] used SD based modelling to address the problem of balance the continuous increase patients and government's targets of patient access performance. In the same line with this, [12] adopted SD model to simulate the demand between paediatric population and paediatric workforce. Then, through the change of some variables, the reasonable number of paediatric workforce was obtained in order to maximize the staff efficiency. Therefore, and considering the fact that patient flow is affected by system bottlenecks which can be relieved by resource reallocation or bed-balancing [14], solutions to these matters are of high priority at health care clinics, in special when all the dynamics from the domain can be considered, since they affect directly the clinical decision-making, thus the service provided to patients. Considering this dynamics, main clinical management problems are still open, such the antenatal clinics patients flow management, which is going to be covered in this research as a contribution to this health care area.

3   Modelling the Antenatal Clinic Decision-Making from a System Dynamics Perspective

To model antenatal clinics, as established by [13], it is important to consider that health care delivery process from the patient's perspective generally is a composition of several care services. In this context, planning and control will highly affect the effectiveness and efficiency of the health care decisions and delivery that are involved in in every care stage. Thus, a proper interview process was conduced at the British Northwest Hospital antenatal clinics which belongs to he NHS TRUST in the UK (due to confidentially, clinics names are not used, but we will use the acronym HANC to represent the British northwest hospital antenatal clinics). From this, it was realised that relationships between patients and medic staff activities are related in many ways and it is relevant to provide a good understanding of this (details for the information collected is presented in Sect. 3.2). Therefore, the SD model, through the us of causal loop diagrams, is considered in order to understand, realise and visualise how all the activities are linked each other. Within this, the main elements and structure for the computer based simulation model will be supported as presented in Sect. 4. Mainly, and based on the modelling HANC requirements, that are considered as relevant for enhancing the business process regarding to the criteria's established by [13], two perspectives for this patient flow are considered. In the first place, the perspective of the patient will represent all the flow and mechanism that patients experience once they attend the HANC and, in the second place, the perspective of the medical staff is covered within the purpose of realize the main activities, and their implications, carried out during the clinic appointment. These perspectives are explained on the following sub-sections.

3.1   A Causal Loop Diagram of Patient Perspective

The following diagram (Fig. 1) shows how the patient factors have influence on the whole process and relationship loops of different elements. These factors from patient perspective are patient attendance rate, pregnancy stage and number of missing forms that would be discussed in details as follows. These elements are basically assumptions collected from the meetings at the HANC's, which are considered relevant for providing a dynamic understanding of what are the relationships across them.

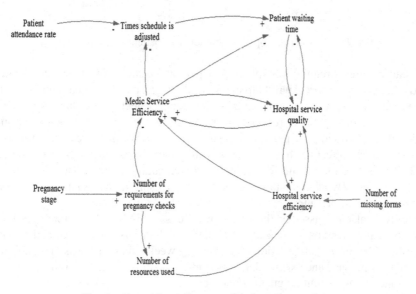

Fig. 1. Causal loop diagram of patient perspective.

As depicted in Fig. 1, it is possible to say that if patients arrives to the hospital on time or not, the impact on the waiting time will be highly affected. This is because once a patient arrives late; the appointment schedule has to be manually adjusted according to the current patient attendance. Therefore, the staff will spend unnecessary time doing these arrangements implying potential increment on the original waiting time for patients who have arrived at waiting area on time. Hence, late patients are managed for being attended as soon as possible and, as a result, the patient waiting time would be increased as well. Finally, the hospital service quality and hospital service efficiency would be further impacted because long waiting time in hospital might bring up the not satisfied patients', which has been mentioned in the key challenges in NHS above. However, the view of patients is one of the most important evaluations in hospital service quality and efficiency. So, if the total waiting time were increased, the elements of quality and efficiency would be reduced (see Fig. 2).

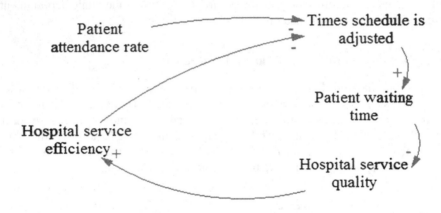

Fig. 2. Reinforcing loop about influence of patient attendance rate.

Because patients require to explain their personal case history when they come to the HANC, the use of forms will help medic staffs to grasp different patients' situations and realize patients' unique requirements like doctor selection and language requirements. Therefore, if patients forget to bring their case history, they would be given additional time to fill in basic personal information, which would increase the processing time for staff and affect the hospital service efficiency. Then, as can be represented in Fig. 3, if there are more time for medics to understand patients' conditions, medics' working efficiency would inevitable be reduced thus resulting in less patients can access to the medics' staff on time.

Finally, as also discussed on the interview process at HANC, the factor of pregnancy stage would also affect on the workflow in antenatal clinics. As mentioned in the process description, the more accumulation of pregnancy weeks the more tests and checks are required to be planned and executed. In this context, medics would generally spend more time to find out the current pregnant situation of patients and require patients to do more checks to reduce the risk of delivery. Therefore, the extension of treatment time would increase the possibility of long waiting (Fig. 4 shows the dynamic causal loop diagram

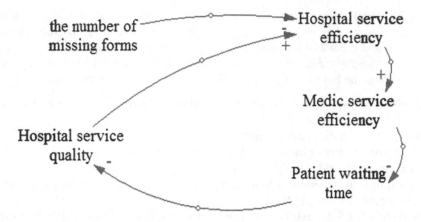

Fig. 3. Balancing loop about influence of missing forms.

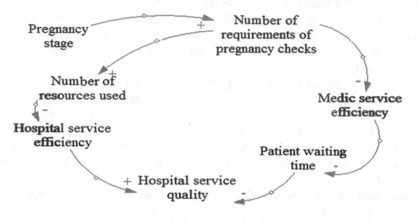

Fig. 4. Reinforcing loop about influence of pregnancy stage.

for this situation). On the other hand, as the number of tests increases, more resources would be used, which would reduce the efficiency of the patient service.

3.2   Data Collection Definition for Visualizing the Patient Flow Behaviour

In order to give reality to this model, data is needed to be collected in order to make rational decisions, evaluate the processes performance, mainly in relation to the management objectives and fulfil, HANC requirements. Therefore, the data will be useful for supporting the definition of key performance indicators that are useful for assessing objectives. In the most of the cases, those indicators must be tailored to each process under study, which are dependent on which social, economic or environmental concerns are important. Appropriate indicators can be developed which measure the state of the resource, the performance controls, economic efficiency, socio-economic performance and social continuity. Therefore, based on this, and considering the HANC clinic requirements', the information to be collected is presented in the following section as well as the results obtained from this process. Therefore, the information to be collected is in order to be used as support, in the first place, for understanding and validating the patient flow and behaviour across the antenatal clinic and, in the second place, in order to be used as input of the simulation model. Therefore, and based on the interviews with member staff at the HANC's, the information to be collected is as follows:

- **Hospital Number**: Unique number from patients in order to track the frequency of attendance to the clinic.
- **Date**: Date when the patient attend the clinic.
- **Clinic Appointment time**: Pre-defined appointment time.
- **Postcode**: origin address of the patient.
- **Arrival time at Antenatal Clinic Reception**: Real time at which the patient arrive to the reception in the HANC.
- **US arrival & US departure**: This is information is required to realise if the patient, before attending the HANC, was already in the Ultrasound area. This information is useful to detect if there is any delay regarding to this.
- **Extra Activity 1, time start & end**: Before the clinic appointment, once the patient arrives, they may perform additional activities required for the clinic. In some cases patients will arrive much earlier for performing this activities, such as the glucose tolerance test.
- **HCA check of blood Pressure and Urine, time start & end**: Before the clinic appointment, nurses provide support to clinic by running ambulatory test that will be useful for clinic appointment. In this case, the HCA of blood pressure plus the Urine test are mandatory before starting the clinic.
- **Extra Activity 2, time start & end:** In some cases, some tests were not taken before the clinic starts or they required to be repeated. Hence, extra activity 2 measures these extra activities.
- **Clinic appointment, time start & end:** This information collects the effective that a patient spends during the clinic.
- **Staff grade for attending the patient:** Since staff at the clinic appoint can be of different grades, categories such as: Consultant, FT1 to ST2 and ST3 to ST7 are considered in order to realise which staff has attended the patient. From this information it will be able to realise the importance
- **Activities at the clinic appointment**: This information is in order to realise the main activities are performed during the clinic appointment. These activities can be: discussions with other doctors, Caesarean booking, portable ultrasound, blood test, and interpreter needed and/or any other. The purpose of this is to realise which are the most common activities and which one impact the most to the attendance time per patient.
- **Blood test, time starts & end**: Sometimes, after the clinic session, additional blood sample are required. This activity, in terms of time is measured in this element.
- **Pharmacy, time start & end:** In the case that the patient will go to the pharmacy before leaving the clinic, the time regarded this activity is collected here.
- **Antenatal clinic checkout time**: This is the real time at which the patient leaves the ANC.

The aforementioned information was collected by 8 Fridays clinic, in total 93 patients were covered and the results are presented in the following section.

4 Simulation

The simulation model for the SD representation (built in Vensim 6®) model was made on the basic of the information from the combination of two completed causal loop diagrams which is showed Fig. 5. This simulation model consider all the variables already defined in Sect. 3 and it is oriented to reflect the variables change dynamism like the hospital service quality and service efficiency. In fact, this model also considers the waiting time and number of medics available in order to reflect the HANC service quality and efficiency. Therefore, there is no big difference in the establishment of the working process in antenatal clinic. Moreover, through this SD simulation based model, different activities' relationship and their impacts on the whole process will be tested, this in order to support HANC clinical users to better realise their patient flow behaviour. For this simulation, considering the same number of patient arrivals, two scenarios are to be considered which are related to realise the patient waiting time considering the availability of medics. This information is extremely helpful for Hospital management, since these values are normally fixed and very difficult to modify. Thus, through this simulation, it will be shown the impact in such modification (see Table 2).

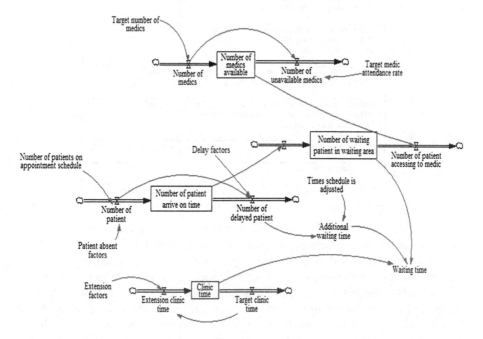

Fig. 5. System dynamic model of workflow in antenatal clinic.

Table 1. System dynamics based decision-making model equations and units.

Time	Equation	Units
unit: Day	Additional waiting time= INTEG (Number of delayed patient*Times schedule is adjusted,0)	minute*person*Day
	Clinic time=Target clinic time + Extension clinic time	minute
	Delay factors= 0.1	person/person [0,0.3]
	Extension clinic time=Target clinic time*Extension factors	minute
	Extension factors= 0.1	minute/minute [0,0.4]
	Number of delayed patient= Number of patient*Delay factors	person
	Number of medics=1*Target number of medics	person
	Number of medics available= Number of medics-Number of unavailable medics	person
	Number of patient= Number of patients on appointment schedule*(1-Patient absent factors)	Person
	Number of patient accessing to medic= Number of medics available	person [?,?,1]
	Number of patient arrive on time= Number of patient-Number of delayed patient	person
	Number of patients on appointment schedule= 50+ (STEP (10,10)-STEP (10,20)+ STEP(10,30) -STEP(10,40)+STEP(10,50))	person
	Number of unavailable medics= Number of medics*(1-Target medic attendance rate)	person [?,?,1]
	Number of waiting patient in waiting area= INTEG (Number of patient arrive on time-Number of patient accessing to medic,1)	person*Day
	Patient absent factors=0.05	person/person [0,0.2]
	Target clinic time=15	minute [10,20]
	Target medic attendance rate=0.8	person/person [0.5,1]

From the system dynamic based decision-making model, it is depicted that all the elements regarding to the number of medics available means the accurate amount of medics who are working during the working period, and is impacted by number of medics and number of unavailable medics. In the mean time, the number of unavailable medics would be changed according to the target medic attendance rate. After that, factor of number of waiting patient in waiting area is affected by number of patient arrive on time and number of patient accessing to medic. Among this relationship, the value of number of patient accessing to medic is consistent with the one of number of medics available because the amount of medics in working time decides the number of patients receiving treatment. Then comes to element of number of patient arrive on time, number of patient and number of delayed patient have influences on it. Simultaneously, number

Table 2. Experiments results.

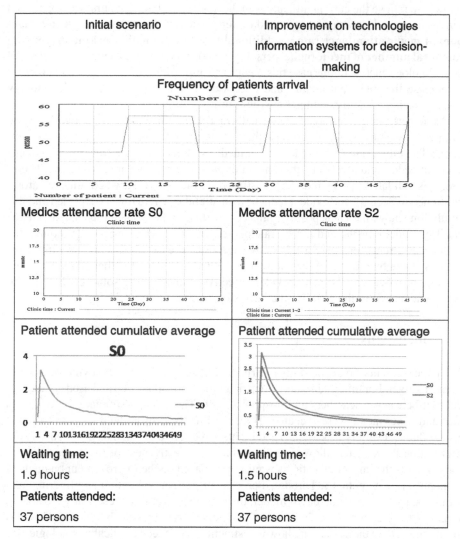

Initial scenario	Improvement on technologies information systems for decision-making

of patient and number of delayed patient would be effected by absent factors like choosing other medical service and delayed factors respectively. The clinic time is also one important element in this model and equal to target clinic time plus extension clinic time. The extension clinic time is impacted by relative extension factors like different pregnancy stage. Finally, the variable of waiting time which can effectively reflect hospital service quality is composed of regular waiting time and additional waiting time. The additional waiting time is caused by readjusting schedule when delayed situations happen. Secondly, value and equation should be set up for every constant and variable. All the equation in the dynamic model, elements' unites and time scale are showed in the Table 1, which are useful to better understand the elements' relationships and basic model's operation.

The experiment one help to test the effectiveness of some improving measures which can help to relieve the current difficulties in the antenatal clinic. The first variable added is investment on the management of medic attendance rate. **Investment on the management of medic attendance rate = additional number of medic working in working time/total number of medic** (unite: person/person). The second experiment, is regarded the technology factor. In this case this means the application of advanced technologies to increase the treatment quality and decrease the clinic time. Where, Technology factor = saving clinic time/clinic time (unit: minute/minute).

As depicted in the variety of elements from Table 2, the number of patient's arrival is quite homogeneous. This is because they normally base their attendance on appointments. Therefore, due to the reduction in medic's attendance rate time, more patients can be attended, and as a result, the number of patient in waiting area is reduced. Therefore, it is validated that by reducing wasting time on unnecessary activities (as aforementioned in Sect. 3), and evident reduction on attendance rate time is expected and this simulation shows the clear positive impacts on this. In fact, since this HANC area is based on appointments, and not on random arrivals (likewise emergency rooms), the number of patients selected for this study are reflect as enough since the overall results remains stable by the end of the simulation. Hence, it is depicted that computer based simulation, based on SD, can provided effective and low cost solutions in order to enhance and support the NHS hospital decision-making.

5   Conclusions

This paper presents a realistic study in British HANC by using system dynamic as main research driver. For this, main causal-loop diagrams were identified and used as communication between researchers and hospital's staff. Therefore, a simulation model was built and experiments were executed. The experiments analyse the responses from the system due to a variety of variables and inputs regarded to the HANC decision-making information flow, which allowed to determine the effectiveness of the proposed solutions, such as the impact on patients waiting rate based on the improvement on medical attendance rate on patients. Through a process of gradual validation and filtering, and within the system dynamic based model, the outcomes from this research were able to suggest strategies for increasing medics' attendance rate time with patients. This was complemented by understanding how investment in advanced medical technologies for enhancing treatment quality, optimisation of system used for hospital staff to increasing efficient working time, setting up an information network for sharing patients' information between neighbouring hospitals and regular staff training could improve patients' satisfaction and hospital service quality. However, limitations of system dynamics such as time length for creating and validating the model, timely modification of activities in the model regarding to new requirements and information inputs, should also be considered by managers during implementation of these improvements in order to avoid unnecessary risks and losses. Further work on this research, in order to better understand the patient flow and their waiting time, will deal with a variety of comparative scenarios by considering rates of: number of patients arriving on time, number of rescheduled patients, rate of patients with more than one medical disorder. Finally, a discrete-event, a multi-criteria and a optimisation decision-making model are also expected to be produced in order to extend this research.

Acknowledgments. This research was supported by Business Research Gateway KE-Voucher funding provided by the University of Liverpool. In addition, visits and questionnaires were supported by the NHS research passport provided to Dr. Hernandez in order to perform this study at British Health Care institutions which also deal with Ethical approval in order to publish these results. In addition to this, authors would like to thanks the support from the Professor Andrew Weeks from the Department of Women's and Children's Health at University of Liverpool, who provided comments and suggestions in order to better align this research activity.

References

1. Brailsford, S.C.: System dynamics: what's in it for healthcare simulation modelers. In: Proceedings of the 2008 Winter Simulation Conference (2008)
2. Chaerul, M.: A system dynamics approach for hospital waste management. Waste Manage. **28**, 442–449 (2008)
3. Homer, J.B., Hirsch, G.B.: System dynamics modeling for public health: background and opportunities. Am. J. Public Health **96**(3), 452–458 (2006)
4. Kuljis, J., et al.: Can health care benefit from modeling and simulation methods in the same way as business and manufacturing has? In: Proceedings of the 2007 Winter Simulation Conference (2007)
5. Lane, D.C., et al.: Looking in the wrong place for health care improvements: a system dynamics study of an accident and emergency department. J. Oper. Res. Soc. **51**, 518–531 (2000)
6. Lincetto, O., et al.: Antenatal Care (2013). http://www.who.int/pmnch/media/publications/aonsectionIII_2.pdf. Accessed 15 Dec 2015
7. Mathai, M.: Alternative Versus Standard Packages Of Antenatal Care For Low-Risk Pregnancy. World Health Organization, Geneva (2011)
8. NHS: Antenatal classes (2015). http://www.nhs.uk/conditions/pregnancy-and-baby/pages/antenatal-classes-pregnant.aspx#close. Accessed 12 Dec 2015
9. Royston, G., et al.: Using system dynamics to help develop and implement policies and programs in healthcare in England. Syst. Dyn. Soc. **15**, 293–313 (1999)
10. Siegel, M., Goldsmith, D.: Improving Health Care Management Through the Use of Dynamic Simulation Modeling and Health Information Systems (2013). http://www.systemdynamics.org/conferences/2011/proceed/papers/P1239.pdf. Accessed 22 July 2013
11. Wolstenholme, J.E.: A case study in commjnity care using systems thinking. J. Oper. Res. Soc. **44**, 925–934 (1993)
12. Wu, M.H., et al.: Population-based study of pediatric sudden death in Taiwan. J. Pediatr. **155**, 870–874 (2009)
13. NICE: Updated NICE guideline on care and support that women should receive during pregnancy. National Institute for Health and Clinical Excellence Report (2013). http://www.nice.org.uk/guidance/CG62/documents/updated-nice-guideline-published-on-care-and-support-that-women-should-receive-during-pregnancy. Accessed 28 Feb 2016
14. Hulshof, P.J., Kortbeek, N., Boucherie, R.J., Hans, E.W., Bakker, P.J.: Taxonomic classification of planning decisions in health care: a structured review of the state of the art in OR/MS. Health syst. **1**(2), 129–175 (2012)

Fuzzy Inference Approach to Uncertainty in Budget Preparation and Execution

Festus Oluseyi Oderanti(✉)

Plymouth Graduate School of Management,
University of Plymouth, Drake Circus,
Plymouth PL4 8AA, UK
Festus.Oderanti@Plymouth.ac.uk

Abstract. In recent times, diverse uncertainties in the global economic environment have made it difficult for most countries to meet their financial obligations. For example, according to statistics from European Commission, 24 out of 29 recorded European Economic Area member countries had budget deficits in 2014. Therefore through modelling and simulations, this paper proposes flexible decision support schemes that could be used in managing the uncertainties in budgeting. Rather than entirely relying on estimates of anticipated revenues (which are uncertain and difficult to predict) in government budgeting, the scheme proposes incorporating fuzzy inference systems (which is able to capture both the present and future uncertainty) in predicting the anticipated revenues and consequently, in proposing government expenditures. The accuracy of fuzzy rule base helps in mitigating adverse effects of uncertainties in budgeting. We illustrated the proposed scheme with a case study which could easily be adapted and implemented in any budgeting scenarios.

Keywords: Decision · Budgeting · Uncertainty · Wage negotiation · Fuzzy logic · Membership · Functions

1 Introduction

The increasing global economic uncertainty has made it difficult in recent years for most countries in the world to cope with their annual budgets. For instance, according to the Organisation for Economic Co-operation and Development (OECD) statistics, 31 out of 38 recorded countries had budget deficit in 2014 [1]. Also, available data on European Commission web sites show that 24 out of 29 recorded European Economic Area (EEA) member countries had budget deficit during same period [2]. Furthermore, the dwindling oil prices in the global market has had some devastating effects on some countries whose sources of revenues mostly depend on petroleum exports such as the OPEC (Organization of the Petroleum Exporting Countries) members. The continuing slump in global oil prices is punching heavy holes in their national treasuries [3]. Many once financially buoyant countries have suddenly result into borrowing in order to meet their financial obligations.

Similarly, the increasing rate of unemployment has made most countries that heavily depend on tax revenues to embark on abrupt and huge budget cuts and borrowing which

© Springer International Publishing Switzerland 2016
S. Liu et al. (Eds.): ICDSST 2016, LNBIP 250, pp. 56–70, 2016.
DOI: 10.1007/978-3-319-32877-5_5

has plunged some into serious debt crisis [4]. According to Broughton [5] despite five years of spending cuts, UK Government deficit is yet to be eliminated. The Chancellor's Autumn Statement in December 2014 implied that overall departmental spending will fall in real terms by about 14 % after 2015–16 and spending commitments in the 2010 parliament have led to inconsistencies and distortions in decision-making [5].

In this study, we examined how fuzzy logic concepts could be applied in budgeting process in order to mitigate the adverse effects of economic uncertainties in predicting government's revenues and respectively/automatically adjusting the corresponding expenditures/spending. We provided a general framework for the scheme and we also demonstrated it using a case study.

2 Budgets and Budgeting Decisions

A government budget is a layout of the government's financial plans on how to spend its revenues [6]. It is a short term plan expressed in financial terms and involves forecasting the nations finance for the following year [7]. National budgets provide the crucial framework for the consistent generation of, and alignment between a nation's resources and its' goals, plans and policies in order to promote economic growth [8]. In order for the Government to spend sustainably, a detail plan on public expenditure is required [9]. With wide-spread impact, the formulation (and therefore decision-making process) of the national budget is crucial in maintaining and promoting a country's overall well-being [8]. However, uncertainty in the future levels of impacting factors can make the forward-planning in budgeting process challenging given the inexact knowledge of future variables and their relation to financial resource availability [10].

According to [11], governments rely heavily on historic data in predicting the future trends of the economy. However, according to Sharma et al. [19] historical records can only represent the past behaviour but may be unable to accurately predict the future behaviour of uncertain factors (such as economic factors used in budgeting). It has been further argued that the classical approach to budgeting does not take into account the uncertainty which may be inherent in the information used in them [12]. A later study suggests that the level of accuracy and application of the predicted future variables can impact the future outcome [13]. Macroeconomic factors such as economic output, unemployment, inflation, savings and investment are key indicators of economic performance and are closely monitored by governments, businesses and consumers. However, these factors are difficult to predict with certainty. Problems that are likely to arise during the budgetary preparation process are: difficulty in forecasting the resource availability, and having unrealistic expectations as a result this [14].

Therefore, fuzzy logic could help in quantifying such uncertain factors and support flexible decisions which can respond to the relative extent of factors affecting them. The 'IF-THEN' modelling of the logic (in which fuzzy sets are applied to rules to decipher outcomes) replicates the human heuristics process (intuitive knowledge and learning from experience) through inference – therefore, enabling advisable decisions to be identified based upon the recognition of the relative importance of different and uncertain variables in a specific problem [15].

2.1 Aim and Objectives

The main aim of this paper is to demonstrate how fuzzy logic concepts could be used in developing decision support systems for modelling the uncertainties inherent in budget preparation and execution decision processes. With the above aim, the specific research objectives are as follows:

- To discuss some of the uncertainties in budgeting processes and illustrate how they could be converted into fuzzy variables using fuzzy sets and membership functions.
- Using Matlab interfaces, to develop prototypes and conceptual fuzzy inference decision models that could be used as decision support schemes in mitigating the adverse effects of uncertain factors in budget preparation and execution.

2.2 Fuzzy Logic and Fuzzy Sets

Risk and uncertainty pervade the whole of economic life, and decisions are constantly having to be made whose outcome cannot be known for certain [16]. Fuzzy logic is a problem solving technique used to deal with vague or imprecise complications. Zadeh [15, p. 338] defines a fuzzy set as "*a class of objects with a continuum of grades of membership*". While traditional risk models used for assessing market, credit, and insurance and trading risk are based on probability and classic set theory, fuzzy logic models are built upon fuzzy set theory and are useful for analysing risks with insufficient knowledge or vague data [17].

Fuzzy logic comprises of four main concepts namely: fuzzy sets, a set defined by some degree of membership [18]; linguistic variables, "variable whose values are words or sentences in a natural or artificial language" Zadeh [16, p. 199]; membership functions, "constraints on the value of a linguistic variable" Dweiri & Kablan [17, p. 717]; and fuzzy if-then rules, rules developed using expert knowledge aiming to describe the relationship between the input and output variables.

2.3 Fuzzy Inference Systems

A fuzzy inference system (FIS) (or fuzzy decision making system (FDMS)) uses a collection of fuzzy membership functions and decision rules that are solicited from experts in the field to reason about data [17, 19]. Typical components of a fuzzy decision making system are as shown in Fig. 1. The components of a FDMS, as shown in the figure are; a fuzzification section, a fuzzy rule base, fuzzy decision logic and defuzzification section [20, 21].

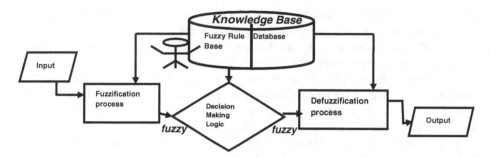

Fig. 1. Fuzzy Inference System (FIS). The figure shows components of a fuzzy decision making system.

3 Overview of the Proposed Approach

The proposed scheme will be termed *"Fuzzy Inference approach to Uncertainty in Budget preparation and Execution"* (FUBE). The proposed approach used fuzzy rules that are solicited from experts to develop flexible decision support scheme that could be used in capturing the uncertain economic factors that are necessary in predicting the expected revenues and in respectively formulating and proposing the expected government expenditures that could help governments to stay within their means (i.e. avoid spending beyond their means/revenues) and avoid perpetual budget deficits that has turned most countries to chronic debtors and thrown many into deep national economic recession.

4 Rationale for the Scheme and Contributions

a. The scheme could assist government in spending within their means and avoid budget deficits.
b. It would help the government in prioritising their spending because as a result of the efficiency of the fuzzy rules from the fuzzy knowledge base, some low-priority expenditures might be cut off from the list if the expected revenues they depend on fail to generate relative income into the budget scheme.
c. It would help government to avoid sudden and unexpected budget cuts because the scheme would automatically allocate expenditure based on generated revenue and this will be prioritised as explained in item b above.
d. The proposed automated decision scheme could help in reducing man hours lost on routine decisions on budget preparation and execution.

5 Assumptions

In this report, we acknowledge that there are numerous types of uncertain factors, revenues and public expenditures that are considered during budgeting. For simplicity, this report will be focusing on four uncertain revenue variables as inputs into the decision systems and these include: inflation, GDP, unemployment and income tax. We are assuming that all other macroeconomic factors remain constant [21]. We have chosen the variables to illustrate our framework for simplicity and because of the interwoven relationships among the variables and their respective outputs.

According to Barnes [22], the relationship between inflation and economic output (GDP) plays out like a very delicate dance. For stock market investors, annual growth in the GDP is vital. If overall economic output is declining or merely holding steady, most companies will not be able to increase their profits, which is the primary driver of stock performance. However, too much GDP growth is also dangerous, as it will most likely come with an increase in inflation, which erodes stock market gains by making money (and future corporate profits) less valuable [22, 23]. But where do these numbers come from? In order to answer that question, we need to bring a new variable, unemployment rate, into play. Studies have shown that over the past 20 years, annual GDP growth over 2.5 % has caused a 0.5 % drop in unemployment for every percentage point over 2.5 % [22, 24].

Furthermore on the assumptions, although it is suggested that expert knowledge should be sought and used when formulating decision rules and membership functions for the knowledge base [17, 25]; this study is using the author's economic and business knowledge to articulate the decision rules, which are used in the proposed decisions scheme. In a real system implementation, the validity or the relationship between the inputs and output variables would also be determined by relevant experts in the field.

6 Uncertain Factors in Budgeting

6.1 Inputs to the Decision Systems

The inputs to the proposed *Fuzzy Inference approach to Uncertainty in Budget preparation and Execution* (FUBE) decision systems are the examples of what normally constitute input factors during budget preparation and these could be in terms of uncertain (predicted) government revenues or other uncertain economic factors that could affect the processes. Some of these input factors are as discussed below.

6.1.1 Inflation

Inflation can be defined as a continuing rise in the level of consumer prices [26]. The nature of the inflation process makes it very difficult to forecast, even when inflationary conditions in the economy appear to be benign. External economic shocks can make forecasts inaccurate. For example, a hike in world oil prices (an inflationary shock) or deep falls in global share prices (a deflationary shock), both have tremendous effects on economic systems [27]. The exchange rate might also fluctuate leading to volatility in the prices of imported goods and services.

In order to calculate inflation (as an input for FUBE) for a particular year, we simply calculate the percentage change of consumer price index (CPI) as follows:

$$\text{Inflation} = \frac{CPI_1 - CPI_0}{CPI_0} * \frac{100}{1}$$

Where CPI_0 is the initial value and CPI_1 is the final value.

6.1.2   Unemployment

The International Labour Organization (ILO) measure of unemployment assesses the number of jobless people who want to work, are available to work and are actively seeking employment [28]. Unemployment is calculated thus:

Unemployment rate $=$ (Unemployed Workers/Total Labour Force) $* 100 \%$

The economic costs of unemployment to a government budget are immense. Unemployment leads to higher payments for unemployment benefits such as unemployment insurance, unemployment compensation, welfare, subsidies, government-funded employment (in extreme cases) and a reduction in the level of income tax generated, forcing the government to borrow money or cut back on other spending [29]. The uncertainty and unpredictability of unemployment is as shown in Fig. 3. The figure illustrates the fluctuation of unemployment and jobseeker's allowance in the UK between 1992 and 2015.

6.1.3   Economic Growth and Gross Domestic Product (GDP)

Economic growth refers to an increase in real GDP. This in turn implies an increase in the value of national output/expenditure [30]. GDP growth can be affected by several factors – interest rates, asset prices, value of exchange rate, political stability etc.

The following statements from the UK Office for Budget Responsibility [11] further buttress the volatility/uncertainty of GDP. According to OBR [11, p. 3] "We forecast that GDP would rise by 5.7 % from the first quarter of 2010 to the second quarter of 2012, but the latest data suggest it has grown by only 0.9 per cent. To begin with we raised our short-term growth forecasts in the autumn of 2010, in response to the unexpected strength of GDP that summer, only to revise them down again as the economy lost momentum going into 2011. We then forecast a broadly flat profile for GDP into and through 2012 in our November 2011 and March 2012 EFOs, only to see GDP fall steadily in the most recent three quarters" [11].

GDP can be calculated thus:

i. **GDP** = CS + CI + GS + NX
   Where CS is Consumer Spending, CI is money spent on investment, inventories equipment etc., GS is total government spending and NX is net exports.
ii. **Percentage rate of growth in GDP** = 100 * X, where X is the solution of VB * (1 + X) ^ N = VE.
   Where VB is value of good or service at the beginning of the valued period, N is number of periods between VB and VE usually number of years and VE is the value of good or service at the end of the valued period.

6.1.4   Tax Revenue

Tax revenue is very crucial to many countries, such as UK, national budget preparation as it is the main source of government finance needed to allocate funds for expenditures [31]. The large array of factors impacting the revenue generated through different types of tax (i.e. income tax, VAT, etc.) means the forecasts of future tax revenues cannot be precise. For example, unemployment levels can affect income tax revenue, and a recession can result in less money being generated by corporation tax on businesses' profits [32].

6.2   Output of the Decision System

The outputs from the *FUBE* decision systems relate to the examples of government expenditures (i.e. spendings) which are mostly affected by input factors that were discussed in Sect. 6. Few of these output factors and their assumed (Sect. 5) relationships with input factors are as discussed below. As stated in Sect. 5, we acknowledge that even though, not all the expenditures could be varied (or subjected to fuzzy fluctuations/variations that are based on variation of the uncertain input factors) and while some are protected, fixed or mandatory. However, expenditures like stipends paid to the unemployed citizens (unemployment benefits/job seeker allowance) could be subjected to such variations since the beneficiaries do not have any job contract (i.e. contractual obligation) with the government and therefore, their monthly payment could be subjected to variations/fluctuations based on government's ability to pay, income or dictates of the economy. This is consistent with the research conducted in [21] in which the authors concluded that wage increase (or decrease) negotiation between employers and employees could be automated using fuzzy sets such that it depends on uncertain factors (such as firm profits, inflation, etc.) that affect firm's operations and ability to pay.

6.2.1   Output: NHS Budget

The UK National Health Service (NHS) provides universal access to comprehensive healthcare, funded by taxation, free at the point of use and it came into existence on the 5th July 1948 [33]. According to Crawford and Stoye [34], recent repeated calls for above-inflation increases in spending on the NHS over the new Parliament arise from concerns over the demand and cost pressures faced by the NHS [34]. The NHS typically faces above-inflation pressures on its budget from rising staff wages and high-cost drugs. NHS England estimates that these factors could amount to pressure on the NHS budget of around 3.5 % per year. A report from UK Royal College of Nursing in [35] reveals that as a result of economic uncertainties, the Government plans to cut £200 million from public health budgets in England under the government's proposal for a blanket 6.2 % funding reduction in January next year and London faces cuts of £5 per person while national average is estimated at £3.77.

According to Appleby [36], during Tony Blair regime, UK was spending 7 % of its GDP on health care – 1.5 % privately and just 5.5 % of public money on the NHS. The EU-15 average spend in 2000 was around 9 % of GDP. Under a range of assumptions, the latest OBR projections suggest NHS spending by 2063/4 of anything between 7.5 % and 14.4 % of GDP [36]. Further relationship between inflation, GDP and UK NHS budget have been explained in many other studies which include [5, 11, 34].

Therefore, in the case study in Sect. 8 of this paper and as shown in Fig. 2, we assumed government spending on NHS (or health sector) as an output that is affected by inflation and GDP growth.

Table 1. Example of uncertain variables (revenues and expenditures)

Year	Unemployment rate	GDP Growth rate	Inflation rate	Tax revenue (Billion £)	Unemployment benefits (Billion £)	Healthcare budget (Billion £)
2015	5.4	0.7	−0.1	515	3.07	134.1
2014	7.2	0.7	1.9	494	4.34	129.2
2013	7.8	0.5	2.5	474	5.17	124.3
2012	8.4	0.1	2.8	472	4.93	121.2
2011	8.1	0.5	4.5	454	4.48	119.8
2010	7.9	0.6	3.3	415	4.68	116.9
2009	6.7	−1.3	2.2	446	2.86	108.7
2008	5.3	−1.9	3.6	456	2.24	102.3

Example of uncertain variables (revenues and expenditures) in budget preparation and executions (sources: [37–39]). The input variables are as shown in the first four columns while the output variables are as shown in the last two variables.

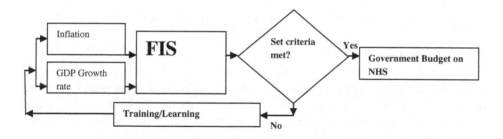

Fig. 2. Proposed fuzzy budget model showing the relationship between the inputs (inflation and GDP) as well as the Output (i.e. Government Budget on NHS)

6.2.2   Output: Unemployment Benefits

Tax revenues (such as income tax) have direct relationship with the budget preparation of most governments. The higher the income tax the less consumers have to spend as disposable income impacting on value added tax [40]. However a heightened income tax leads to more revenue which leads to budget surplus. Income tax is dependable on employment rates which are continually fluctuating as lower unemployment reduces the amount of income taxable. When income tax is low and unemployment is low the government will be receiving low revenues as many people are out of work, affecting their disposable income and affecting the amount of people that are eligible to be taxed upon therefore there will be less money to spend on unemployment benefits (also known as Jobseeker's Allowance).

Therefore, for any government to plan for spending on unemployment benefits, it needs to consider factors such as unemployment rates and tax revenues. Figure 3 illustrates the relationships between Unemployment and Jobseeker's Allowance. Several studies [41, 42] have found that taxation is a significant factor when explaining differences in unemployment rates and hence, the spending on unemployment benefits. Therefore, in the case study (Sect. 8) of this paper and as shown in Table 1, we assumed that government spending on unemployment benefits as an output that is affected by tax revenues and unemployment rate/data.

The relationship between the two variables and government spending on benefits is further illustrated in Table 1. In 2011 where unemployment rose, and income tax drops it resulted in increased government spending on benefits. The relationship between the variables and the output are examined further in the Sect. 8 of this report.

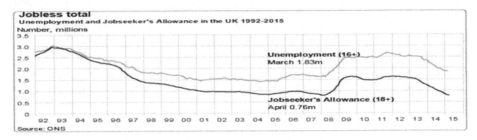

Fig. 3. Unemployment and Jobseeker's Allowance in the UK 1992–2015 (Source: [43]).

7 Methodology

The proposed decision system is divided into two modules for the assumed two outputs (expenditures) respectively. One of the modules is as shown in Fig. 2. Following the research methodologies proposed in [17, 20], to use fuzzy decision making for budget preparation and execution (FUBE) the following procedure is proposed:

a. List all uncertain (fuzzy) factors that will be considered in taking the budgeting decisions: As explained in Sect. 6, we will be choosing six variables of inflation, GDP, unemployment, tax revenues, government spending on health as well as government spending on unemployment benefits. We have chosen these variables to illustrate our framework for simplicity and because of the interwoven relationships among the variables.

b. Determine the input and output variables of the FIS: The proposed decision system is divided into two modules each with two inputs and one assumed output (expenditure) respectively. One of the modules is as shown in Fig. 2. The first module has inputs as unemployment and income tax with its Output as government expenditure on Unemployment benefits. The second module has the inputs as inflation and GDP with the output as Government Budget on healthcare sector.

c. Develop fuzzy sets, subsets and membership functions for all the input and output variables. This can be accomplished by soliciting knowledge from the experts or searching through literature data. Examples of our adopted fuzzy sets, subsets and membership functions are as shown in Fig. 4.

d. Formulate decision rules for the rule base. The rules also ought to be solicited from experts [17]. The rules shown in Fig. 4 depict our adopted decision rules for the knowledge base of the fuzzy inference systems (FIS).

e. Establish relationships between input values and their fuzzy sets and apply the decision rules using the relationships shown in Fig. 4. The fuzzy rule base was coded into a Fuzzy Inference System (FIS) using the Matlab toolbox. Sample rules of the fuzzy inference system are shown in Fig. 4.

f. Use the fuzzy inference system (FIS): Using Matlab fuzzy toolbox, all the fuzzy inputs are passed into the Mamdani type FIS.

g. Get the defuzzified output from the FIS: The crisp output for the **FUBE** is computed using centre of gravity method (COG) method. Figure 5 shows the FIS interface for the output variable *Unemployment Benefit* and a defuzzified (crisp) output interface is as shown in the figure.

h. Training and performance evaluation: Training (or learning) of the FUBE decision model was accomplished through the optimization of the fuzzy logic parameters while using the adaptation of the fuzzy membership functions as the basis for the performance measure as in [20]. This training or learning of the membership functions to optimize its performance was achieved through the use of the *Nelder-Mead* Simplex Search Method for finding the local minimum x of an unconstrained multivariable function $f(x)$ using a derivative-free method and starting at an initial estimate. Further information on this training procedure is available in [45].

8 FUBE Case Study and Results

As an illustration of a case study of the decision support system, we will be using UK budget data from Table 1. The table shows the data for inflation, GDP, unemployment and tax revenues and government expenditure on healthcare and unemployment benefits. Following the **FUBE** methodology explained in Sect. 7, we will pass the uncertain economic factors into the **FUBE** decision systems as inputs (i.e. inflation, GDP, unemployment and tax revenues) to produce the outputs which demonstrate what should have been the *expected or suggested* government expenditures/spending for those years on healthcare and unemployment benefits.

The decision system for the UK case study has been divided into two modules. The first module has its inputs as unemployment rate and government tax revenues while the output is expected/suggested UK Government budget on Unemployment benefits. The second module inputs are inflation and GDP growth rate with the output as expected/suggested Government Budget on health. In this case study, we will be using data from the last eight years to develop the fuzzy rules and demonstrate what should have been the government expenditures (FUBE outputs) for those eight years under consideration if the government had used FUBE decision scheme.

Implementing the proposed decision system with fuzzy toolbox in Matlab, Fig. 4 shows the Mandani defuzification interface and the membership functions for Inflation as an input variable. Also, Fig. 5 shows the results (outputs) of the decision system which represent what would have been the *expected or suggested* Government spending for the years under consideration if the Government had incorporated the FUBE decision system in their budget modelling.

Table 2. Outputs of the FUBE decision system (columns 4 & 5) given the uncertain economic factors

Year	(Actual Govt. Spending) Unemployment benefits (Billion £)	(Actual Govt. Spending) Healthcare budget (Billion £)	(FUBE Output) Expected unemployment benefits (Billion £)	(FUBE Output) Expected healthcare budget (Billion £)
2015	3.07	134.1	4.1	129.0
2014	4.34	129.2	5.0	125.0
2013	5.17	124.3	3.8	115.0
2012	4.93	121.2	3.6	115.0
2011	4.48	119.8	3.8	118.0
2010	4.68	116.9	2.6	112.0
2009	2.86	108.7	3.2	105.0
2008	2.24	102.3	3.0	105.0

The table shows the outputs of the FUBE decision system in columns four and five given the economic uncertain factors (as inputs from columns two to five of Table 1) in comparison with actual spending by Government (column two and three of the table) during same years. Columns four and five represent what would have been the expected Government spending for the years under consideration if the Government had incorporated FUBE decision system in the budget modelling.

9   Results Discussion

Using the procedures highlighted in Sect. 7 above, the results of the FUBE UK case study (shown in columns 5 and 6 of Table 2) show that the flexible decision scheme was able to reduce government expenditures in most of the years under consideration (with respect (or subject) to the uncertain input factors) when compared with the actual expenditures shown in Table 1. This was made possible by the intelligent fuzzy rules from the FUBE knowledge base which is able to effectively capture the present and future uncertainties inherent in budget preparation and execution.

Therefore, we infer that had such intelligent decision system been integrated into budgeting process in the past, it could have helped governments to reduce the menace of perpetual budget deficits and public debts [45] as well as in avoiding unprepared budget cuts [46, 47].

As stated under our assumption in Sect. 5, we agreed that there are numerous factors that are taken into consideration during (or that affect) budget preparation and

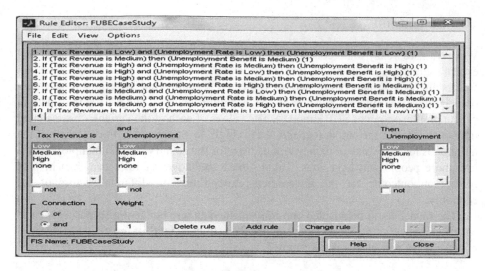

Fig. 4. Examples of fuzzy rules for the FUBE decision systems with the inputs (Tax revenue and Unemployment rate) and the output (Government spending on unemployment benefits).

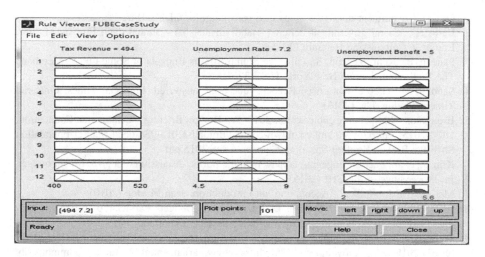

Fig. 5. Sample defuzzification output showing that in year 2014 when Tax revenue was £494billions and Unemployment rate was 7.2, the expected suggested UK Government spending on Unemployment benefits ought to be £5billions as shown in Table 2.

execution, however, when implementing the proposed system in the real environment, these additional factors would be entered into the decision system as additional inputs and fuzzy rules would be developed to model them accordingly.

After the training (learning) of the decision system and in accordance with Sect. 7 step h, the FUBE decision model performs better and this shows that the learning is important as the model is able to adapt with fuzzy reasoning over time [20, 44].

10 Conclusion

We have used fuzzy logic concepts to model uncertainties in budget preparation and execution to create models with fuzzy rules which can be used for capturing uncertain factors in budget preparation and in respectively estimating government spending (expenditures in different sectors of the economy) during budget implementation/execution (i.e. output of FUBE model). The values of each of the variables have a great impact on whether the output (expenditure) is high, low or medium. For example, if the inflation is low the value of money will be high for the country, and if unemployment is low more people will have jobs. Taking these as budgeting rules can then help to predict an outcome for the government expenditures in different sectors of the economy.

The model could be adapted and implemented in any budgeting scenarios such as in setting budget of business organisations, departments, units of organisations, project managements and other related settings.

References

1. OECD, Government deficit / surplus as a percentage of GDP 2014/1
2. Eurostat, General government deficit (–) and surplus (+) - annual data, Eurostat, Editor. European Commission Belgium (2015)
3. Faucon, B., Kent, S., Said, S.: Oil-Price Slump Strains Budgets of Some OPEC Members. The Wall Street Journal New York, U.S. (2014)
4. Sandbu, M.: Greece debt crisis: Readers' questions answered. Financial Times. Financial Times, London, UK (2015)
5. Broughton, N.: Spending choices after 2015; Pre-Budget Briefing, Socialmarket Foundation, UK (2015). http://www.smf.co.uk/wp-content/uploads/2015/03/Social-Market-Foundation SMF-Pre-Budget-Briefing-Spending-Choices-After-2015.pdf
6. Kim, J., et al.: Factful: Engaging taxpayers in the public discussion of a government budget. In: Proceedings of the CHI (2015)
7. McLaney, E.J., Atrill, P.: Accounting: an introduction. Jean Morton (2010)
8. IBP. Why are budgets important. International Budget Partnership (2015). http://internationalbudget.org/getting-started/why-are-budgets-important/
9. Webb, D.: Sustainable public spending. Key Issues for the New Parliament 2010 (2010). [cited 2015 23rd November 2015] http://www.parliament.uk/documents/commons/lib/research/key_issues/Key-Issues-Sustainable-public-spending.pdf
10. Todorovic, J.D., Djordjevic, M.: The importance of public expenditure management in modern budget systems. Econ. Organ. **6**(2), 14 (2009)
11. OBR, Forecast evaluation report. Office for Budget Responsibility, London (2012)
12. Kuchta, D.: Fuzzy capital budgeting. Fuzzy Sets Syst. **111**(3), 367–385 (2000)
13. Karanovic, G., Gjosevska, B.: Application of fuzzy logic in determining cost of capital for the capital budgeting process. Procedia Econ. Finan. **3**, 78–83 (2012)
14. Council, C.: Budget Preparation (2013)
15. Zadeh, L.A.: Fuzzy sets. Inf. Control **8**(3), 338–353 (1965)
16. Zadeh, L.A.: The concept of a linguistic variable and its application to approximate reasoning—I. Inf. Sci. **8**(3), 199–249 (1975)

17. Dweiri, F., Kablan, M.: Using fuzzy decision making for the evaluation of the project management internal efficiency. Decis. Support Syst. **42**(2), 712–726 (2006)
18. Bojadziev, G., Bojadziev, M.: Fuzzy Logic for Business, Finance, and Management. World Scientific Publishing Co., Inc., Singapore (2007)
19. Sharma, R.K., Kumar, D., Kumar, P.: FM – a pragmatic tool to model, analyse and predict complex behaviour of industrial systems. Eng. Comput. **24**(4), 319–346 (2007)
20. Oderanti, F.O., De Wilde, P.: Dynamics of business games with management of fuzzy rules for decision making. Int. J. Prod. Econ. **128**(1), 96–109 (2010)
21. Oderanti, F.O., Li, F., De Wilde, P.: Application of strategic fuzzy games to wage increase negotiation and decision problems. Expert Syst. Appl. **39**(12), 11103–11114 (2012)
22. Barnes, R.: The Importance of Inflation and GDP, in Investopedia. Investopedia, USA (2016)
23. Kilian, L., Hicks, B.: Did Unexpectedly Strong Economic Growth Cause the Oil Price Shock of 2003–2008? J. Forecast. **32**(5), 385–394 (2013)
24. Daly, M.C., et al.: A search and matching approach to labor markets: did the natural rate of unemployment rise? J. Econ. Perspect. **26**(3), 3–26 (2012)
25. Oderanti, F.O.: Fuzzy inference game approach to uncertainty in business decisions and market competitions. SpringerPlus **2**(1), 1–16 (2013)
26. Bucks, B.K., et al.: Changes in US family finances from 2004 to 2007: evidence from the survey of consumer finances. Fed. Res. Bull. **A1**, 95 (2009)
27. Cologni, A., Manera, M.: Oil prices, inflation and interest rates in a structural cointegrated VAR model for the G-7 countries. Energy Econ. **30**(3), 856–888 (2008)
28. Wieland, V.: Monetary policy and uncertainty about the natural unemployment rate (2002)
29. Subhaashree, S.: Application of Fuzzy Logic to Unemployment Problem. Masters Dissertation, Guide: Dr. WB Vasantha Kandasamy, Department of Mathematics, Indian Institute of Technology (2001)
30. Reinhart, C.M., Rogoff, K.S.: Growth in a time of debt (digest summary). Am. Econ. Rev. **100**(2), 573–578 (2010)
31. Piana, V.: Tax Revenue (2003), [cited 2015 27th November, 2015]. http://www.economicswebinstitute.org/glossary/taxrev.htm
32. Pettinger, T.: Tax Revenue Sources in UK (2014) [cited 2015 27th November, 2015]
33. Bloor, K.: Radicaliism and reality in the national health service: fifty years and more. In: Bloor, K. (ed.) Center for Health Economics, the University of York, UK: York, UK (1998)
34. Crawford, R., Stoye, G.: The outlook for public spending on the National Health Service. Lancet **385**(9974), 1155–1156 (2015)
35. Griffin, M.: Cuts in public health funding will hit areas of greater need hardest. Nurs. Stan. **29**(51), 11 (2015)
36. Appleby, J.: How much should we spend on health care? Surgeon **13**(3), 121–126 (2015)
37. Chantrill, C.: Total Public Spending (2015), [cited 2015 5th November 2015]. http://www.ukpublicspending.co.uk/total_spending_2008UKbn
38. Trading Economics. United Kingdom Unemployment Rate 1971–2015 (2015), [cited 2015 5th November, 2015]. http://www.tradingeconomics.com/united-kingdom/unemployment-rate
39. DWP. Official Statistics Benefit expenditure and caseload tables 2015 (2015), [cited 2015 5th December 2015]. https://www.gov.uk/government/statistics/benefit-expenditure-and-caseload-tables-2015
40. Lipsey, R.G., Steiner, P.O., Purvis, D.D.: Economics, 8th edn. Harper and Row Publishers, New York (1987)
41. Nickell, S.: Unemployment and labor market rigidities: Europe versus North America. J. Econ. Perspect. **11**, 55–74 (1997)

42. Scarpetta, S.: Assessing the role of labour market policies and institutional settings on unemployment: a cross-country study. OECD Econ. Stud. **26**(1), 43–98 (1996)
43. BBC. Economy tracker: Unemployment. News Business (2015) [cited 2015 30th November, 2015]
44. Sarakhsi, M.K., Ghomi, S.F., Karimi, B.: A new hybrid algorithm of scatter search and Nelder-Mead algorithms to optimize joint economic lot sizing problem. J. Comput. Appl. Math. **292**, 387–401 (2016)
45. Ostro, Z.K.: In the debt we trust: the unconstitutionality of defaulting on american financial obligations, and the political implications of their perpetual validity. Harv. J. Legis. **51**, 241 (2014)
46. Hamnett, C.: Shrinking the welfare state: the structure, geography and impact of British government benefit cuts. Trans. Inst. Br. Geogr. **39**(4), 490–503 (2014)
47. O'Hara, M.: Austerity Bites: A Journey to the Sharp End of Cuts in the UK. Policy Press (2015)

DSS to Support Business Resilience/Risk Management and Project Portfolio Management

Detectability Based Prioritization
of Interdependent Supply Chain Risks

Abroon Qazi[1(✉)], John Quigley[1], Alex Dickson[1], Şule Önsel Ekici[2],
and Barbara Gaudenzi[3]

[1] Strathclyde Business School, University of Strathclyde, Glasgow, UK
{abroon.qazi,j.quigley,alex.dickson}@strath.ac.uk
[2] Industrial Engineering Department, Dogus University, Istanbul, Turkey
sonsel@dogus.edu.tr
[3] Faculty of Business Economics, University of Verona, Verona, Italy
barbara.gaudenzi@univr.it

Abstract. Supply chain risks must be assessed in relation to the complex interdependent interaction between these risks. Generally, risk registers are used for assessing the importance of risks that treat risks in silo and fail to capture the systemic relationships. Limited studies have focused on assessing supply chain risks within the interdependent network setting. We adapt the detectability feature from the Failure Modes and Effects Analysis (FMEA) and integrate it within the theoretically grounded framework of Bayesian Belief Networks (BBNs) for prioritizing supply chain risks. Detectability represents the effectiveness of early warning system in detecting a risk before its complete realization. We introduce two new risk measures and a process for prioritizing risks within a probabilistic network of interacting risks. We demonstrate application of our method through a simple example and compare results of different ranking schemes treating risks as independent or interdependent factors. The results clearly reveal importance of considering interdependency between risks and incorporating detectability within the modelling framework.

Keywords: Supply chain risks · Risk registers · Systemic · Detectability · Failure modes and effects analysis · Bayesian belief networks

1 Introduction

Supply chains have become complex because of the globalization and outsourcing in manufacturing industries. Global sourcing and lean operations are the main drivers of supply chain disruptions [1]. In addition to the network configuration based complexity, non-linear interactions between complex chains of risks categorized as 'systemicity' of risks [2] make it a daunting task to understand and manage these dynamics. Supply chain risk management (SCRM) is an active area of research that deals with the overall management of risks ranging across the entire spectrum of the supply chain including external risk factors. Besides the increase in the frequency of disruptions, supply chains are more susceptible because of the increasing interdependency between supply chain actors and humungous impact of cascading events [3]. Risk assessment

© Springer International Publishing Switzerland 2016
S. Liu et al. (Eds.): ICDSST 2016, LNBIP 250, pp. 73–87, 2016.
DOI: 10.1007/978-3-319-32877-5_6

comprises three main stages of risk identification, risk analysis and risk evaluation [4]. Generally, risk registers are used in managing risks that treat risks as independent factors [2, 5]. Limited studies have focused on exploring causal interactions between supply chain risks [6, 7]. However, no attempt has been made to capture the detectability associated with each risk within an interdependent network setting of interacting risks. Detectability is an important parameter of Failure Modes and Effects Analysis (FMEA) that represents the effectiveness of an early warning system in detecting a risk before its complete activation [8]. In case of risks having comparable values of the probability and impact, due attention should be given to the risk with lower chance of detection as substantial loss would have resulted by the time it gets noticed. In this study, we focus on the risk analysis stage of risk assessment and propose a new method of prioritizing supply chain risks through integrating the loss and detectability values of risks within the theoretically grounded framework of Bayesian Belief Networks (BBNs) encompassing complex probabilistic interactions between risks.

FMEA is a systematic approach of identifying different modes of failure and evaluating associated risks during the development stage of a product or service. It is known to have been implemented in 1963 for projects at NASA and later, the Ford utilized the technique in 1977 [9]. There are major shortcomings of using Risk Priority Number (RPN) as a measure of prioritizing risks that represents the product of occurrence, severity and detectability associated with each risk [9, 10]. Furthermore, risks are treated as independent factors in FMEA. We adapt the notion of detectability from the FMEA in our modelling framework.

BBN is an acyclic directed graphical model comprising nodes representing uncertain variables and arcs indicating causal relationships between variables whereas the strength of dependency is represented by the conditional probability values [11]. They offer a unique feature of modelling risks combining both the statistical data and subjective judgment in case of non-availability of data [12]. In the last years, BBNs have started gaining the interest of researchers in modelling supply chain risks [7].

1.1 Research Problem and Contribution

In this study, we aim to address the decision problem of prioritizing supply chain risks considering the losses and detectability of such risks within an interconnected probabilistic network setting of interacting risks. Our proposed method contributes to the literature of SCRM in terms of introducing detectability based prioritization of interdependent supply chain risks that has never been explored.

1.2 Outline

A brief overview of the research conducted in SCRM is presented in Sect. 2. The modelling approach of prioritizing supply chain risks is described in Sect. 3 and demonstrated through an illustrative example in Sect. 4. Results corresponding to different ranking schemes and managerial implications are also described in the same section. Finally, conclusions and future research directions are presented in Sect. 5.

2 Literature Review

SCRM is defined as *"the management of supply chain risks through coordination or collaboration among the supply chain partners so as to ensure profitability and continuity"* [13]. Supply chain risks can be viewed with respect to three broad perspectives; a *'butterfly'* concept that segregates the causes, risk events and the ultimate impact, the categorization of risks with respect to the resulting impact in terms of delays and disruptions and network based classification in terms of local-and-global causes and local-and-global effects [14].

Bradley [15] proposed a new risk measurement and prioritization method to account for the characteristics of rare risks contributing to supply chain disruptions. The notion of detection was also incorporated within the model. Segismundo and Miguel [16] introduced a new FMEA based method of managing technical risks to optimize the decision making process in new product development.

Nepal and Yadav [10] presented a methodology for supplier selection in a global sourcing environment and used the techniques of BBNs and FMEA to quantify the risks associated with multiple cost factors. They also introduced rule-based evaluation of risk levels corresponding to 125 different combinations of severity, occurrence and detectability based linguistic variables. Tuncel and Alpan [17] used a timed petri nets framework to model and analyse a supply chain which was subject to various risks. They used the FMEA to identify important risks having higher values of RPN. The major shortcoming of these studies is treating risks as independent factors and/or prioritizing risks on the basis of RPN and related ordinal scales of occurrence, severity and detectability. Furthermore, these studies have not captured the holistic impact of interdependent risks.

In a recent study conducted by Garvey et al. [6], supply chain process and risks corresponding to various segments of the supply network are combined together and modelled as a BBN. New risk measures are also proposed for identification of important elements within the supply network. Their proposed modelling framework differs from the existing BBN based studies in SCRM [7, 18–22] in terms of exploring the propagation impact of risks across the descendant nodes. They also incorporate the loss values within their modelling framework in order to overcome the major limitation of earlier studies that focused exclusively on the probabilistic interdependency between risks without capturing the relative impact of each risk. Qazi et al. [23] introduced new risk measures for capturing the relative impact of each risk on the entire network of interacting risks and proposed methods for selecting optimal combinations of risk mitigation strategies [24] and redundancy strategies [25] keeping in view the risk appetite of the decision maker. However, to the best of authors' knowledge, no attempt has been made to model the detectability of risks within the network setting of systemic risks.

BBNs present a useful technique of capturing interaction between risk events and performance measures [7]. Another advantage of using BBNs for modelling supply chain risks is the ability of back propagation that helps in determining the probability of an event that may not be observed directly. They provide a clear graphical structure that most people find intuitive to understand. Besides, it becomes possible to conduct flexible inference based on partial observations which allows for reasoning [26]. Another important feature of using BBNs is to conduct what-if scenarios [27]. There are certain problems associated with the use of BBNs: along with the increase in number of nodes representing supply chain risks, a considerable amount of data is required in populating the network with (conditional) probability values. Similarly, there are computational challenges associated with the increase in number of nodes.

3 Proposed Modelling Approach

Based on the efficacy of BBNs in capturing interdependency between risks, we consider BBN based modeling of supply chain risks as an effective approach. Such a modeling technique can help managers visualize dynamics between supply chain risks and adopt holistic approach towards managing risks [12, 28]. BBNs have already been explored in the literature of SCRM, however, our proposed BBN based modelling approach is unique in terms of integrating the probabilistic interdependency between risks and loss and detectability associated with each risk within the network setting of interacting supply chain risks.

3.1 Assumptions

Our model is based on following assumptions:

1. Supply chain risks and corresponding risk sources are known and these can be modelled as an acyclic directed graph which is an important requirement of adopting the BBN based modelling approach.
2. All risks are represented by binary states. Instead of focusing on a continuous range of risk levels, we assume that either a risk happens or there is a normal flow of activities/processes (condition of no risk).
3. Conditional probability values for the risks and associated loss and detectability values can be elicited from the stakeholders and the resulting BBN represents close approximation to the actual perceived risks and their interdependency.

3.2 Supply Chain Risk Network

A discrete supply chain risk network $RN = (X_R, G, P, L, D)$ is a five-tuple comprising:

- a directed acyclic graph (DAG), $G = (V, E)$, with nodes, V, representing a set of discrete supply chain risks and risk sources, $X_R = \{X_{R_1}, \ldots, X_{R_n}\}$, loss functions, L, and detectability weighted loss functions, D and directed links, E, encoding dependence relations

- a set of conditional probability distributions, P, containing a distribution, $P(X_{R_i}|X_{pa(R_i)})$, for each risk, X_{R_i}
- a set of loss functions, L, containing one loss function, $l(X_{R_i})$, for each node X_{R_i}
- a set of detectability weighted loss functions, D, containing one detectability weighted loss function, $d(X_{R_i})$, for each node X_{R_i}.

The prior marginal of a supply chain risk or risk source, X_{R_i}, is given by:

$$P(X_{R_i}) = \sum\nolimits_{Y \in X_R \setminus \{X_{R_i}\}} P(X_R)$$
$$= \sum\nolimits_{Y \in X_R \setminus \{X_{R_i}\}} \prod\nolimits_{X_{R_i} \in X_R} P(X_{R_i}|X_{pa(R_i)}) \tag{1}$$

Risk network expected loss, $RNEL(X)$, representing the expected loss across the entire network of interacting risks is an important parameter for assessing the risk level of the supply network under a given configuration of risk mitigation strategies at a specific time. $RNEL(X)$ is the summation of expected loss values across all the risk nodes as follows:

$$RNEL(X) = \sum\nolimits_{X_R} P(X_{R_i})l(X_{R_i}) \tag{2}$$

Risk network expected detectability weighted impact, $RNEDI(X)$, integrates the detectability feature of each risk within the framework of assessing risk level of the supply network under a given configuration of risk mitigation strategies at a specific time. $RNEDI(X)$ is the summation of expected detectability weighted loss values across all the risk nodes as follows:

$$RNEDI(X) = \sum\nolimits_{X_R} P(X_{R_i})d(X_{R_i}) \tag{3}$$

Risk Measures. We introduce two risk measures namely Risk network expected loss propagation measure $(RNELPM)$ and Risk network expected detectability weighted impact measure $(RNEDIM)$ in order to evaluate the relative contribution of each supply chain risk towards the propagation of loss across the entire network of risks. The major contribution of these risk measures is their merit of capturing the network-wide propagation of impact across the web of interconnected risks. The later risk measure is superior to the former in terms of assigning detectability value to each risk and modelling the efficacy of early warning system(s) in detecting risks.

$RNELPM$ is the relative contribution of each risk factor to the propagation of loss across the entire network of supply chain risks given the scenario that the specific risk has happened.

$$RNELPM_{X_{R_i}} = RNEL(X|X_{R_i} = true) * P(X_{R_i} = true) \tag{4}$$

RNEDIM is the relative contribution of each risk factor to the propagation of detectability weighted impact across the entire network of supply chain risks given the scenario that the specific risk has happened.

$$RNEDIM_{X_{R_i}} = RNEDI(X|X_{R_i} = true) * P(X_{R_i} = true) \tag{5}$$

Detectability Scale. Considering detectability as an important parameter in our modelling framework, we follow the detectability scale as shown in Table 1. A detectability value represents the manageability of each risk during the timeframe between reception of early warning signal and complete realization of the risk.

Table 1. Detectability scale [8]

Detectability value	Description
9 or 10	There is no detection method available or known that will provide an alert with enough time to plan for a contingency.
7 or 8	Detection method is unproven or unreliable; or effectiveness of detection method is unknown to detect in time.
5 or 6	Detection method has medium effectiveness.
3 or 4	Detection method has moderately high effectiveness.
1 or 2	Detection method is highly effective and it is almost certain that the risk will be detected with adequate time.

3.3   Modelling Process

Our proposed method comprises three main stages namely problem structuring, instantiation and inference.

Problem Structuring. Supply chain risks and risk sources are identified through involving all the stakeholders of the supply chain. The network structure is developed through connecting the arcs across related risk sources and risks and all nodes are expressed as statistical variables. The problem owner needs to ensure that the model is developed to represent the actual interdependency between risks. The model builder can assist in structuring the model keeping in view the mechanics of a BBN [29].

Instantiation. This stage involves evaluation of (conditional) probabilities either through elicitation from the experts or extraction from the data. Loss and detectability values are also elicited. Probability elicitation is the most difficult task of the modelling process as the experts find it challenging to describe the conditional probabilities.

Inference. In this stage, key risks are identified through measuring the *RNELPM* and *RNEDIM* for each risk and risk source. The beliefs can also be propagated easily once the information on any risk or combination of risks becomes available.

4 An Illustrative Example

4.1 Assumed Risk Network and Modelling Parameters

We demonstrate application of our proposed method through adapting a simple supply chain risk network [6] as shown in Fig. 1. The model was developed in GeNIe [30]. The supply network comprises a raw material source, two manufacturers, a warehouse and retailer. Risks and risk sources are represented by nodes (shown as bar charts) with values reflecting the updated marginal probabilities. Each risk is represented by binary states of True (T) and False (F). Assumed loss values and detectability scores are shown in Table 2 and (conditional) probability values of the risks are depicted in Table 3. These parameters specific to a real case study can be elicited from experts through conducting interviews and focus group sessions.

Table 2. Loss values and detectability scores for the supply chain risks

Supply chain element	Risk	Risk ID	Loss	Detectability
Raw material source (RM)	Contamination	R1	200	1
	Delay in shipment	R2	100	1
Manufacturer-I (M1)	Machine failure	R4	350	1
	Delay in shipment	R5	100	2
Manufacturer-II (M2)	Machine failure	R3	400	2
	Delay in shipment	R6	150	1
Warehouse (W)	Overburdened employee	R7	150	10
	Damage to inventory	R8	100	9
	Delay in shipment	R9	100	8
	Flood	R12	50	10
Warehouse to retailer (W-R)	Truck accident	R10	200	9
Retailer (R)	Inventory shortage	R11	50	2

4.2 Results and Analysis

Once the model was developed and populated with all the parameters, it was updated for obtaining the marginal probabilities of interdependent risks. We analyzed risks with respect to five ranking schemes in order to appreciate the underlying differences. The five schemes can be categorized into two main streams; conventional methods relying on risk registers and the FMEA based tools treating risks as independent factors and the ranking schemes based on our proposed risk measures considering the holistic inter-action between interdependent risks. The updated marginal probabilities were used in assessing risks for all the five schemes. However, in case of methods assuming risks as independent factors, the probabilities are assigned to individual risks without modelling the risk network. Risk measures were calculated for each risk in order to prioritize risks against the three schemes relying on our proposed approach. Finally, we compared the results of all these schemes in order to understand the extent of variation.

Conventional Method of Prioritizing Supply Chain Risks (Scheme 1). As conventional methods rely on the assumption that risks are independent of each other, the resulting ranking profile does not capture interdependency between the risks. Considering

Table 3. (Conditional) probability values: $P(risk = F|parents) = 1 - P(risk = T|parents)$

Parents				P(risk = True|parents)					
R1	R2	R3	R4	R1	R2	R3	R4	R5	R6
				0.3					
	T				0.8				
	F				0.3				
						0.2			
							0.3		
	T	T						0.7	
	T	F						0.4	
	F	T						0.6	
	F	F						0.1	
		T	T						0.9
		T	F						0.6
		F	T						0.5
		F	F						0.2

Parents							P(risk = True|parents)					
R5	R6	R7	R8	R9	R10	R12	R7	R8	R9	R10	R11	R12
							0.3					
		T				T		0.95				
		T				F		0.6				
		F				T		0.8				
		F				F		0.1				
T	T	T							0.9			
T	T	F							0.4			
T	F	T							0.8			
T	F	F							0.3			
F	T	T							0.8			
F	T	F							0.3			
F	F	T							0.7			
F	F	F							0.1			
										0.3		
				T	T						0.9	
				T	F						0.7	
				F	T						0.8	
				F	F						0.1	
												0.2

the two important factors of probability and impact associated with each risk, we prioritized the risks as shown in Fig. 2. The curve represents the risk level (product of probability and impact) of the most critical risk namely R4. However, the result would only be valid in case of all risks modelled as independent factors with no arc connected across any pair of risks. Similar ranking scheme is adopted in most of the ranking schemes based on risk registers.

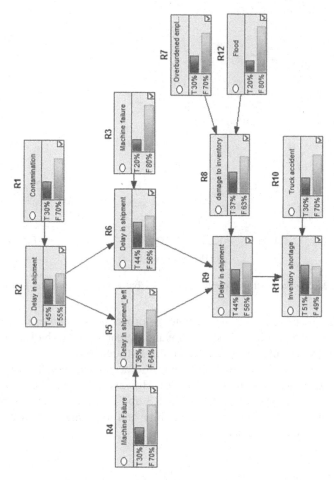

Fig. 1. Bayesian network based model of a supply network (adapted from Garvey et al. [6])

FMEA Based Prioritization of Supply Chain Risks (Scheme 2). In FMEA based ranking schemes, risks are prioritized with regard to the relevant product of occurrence, severity and detectability scores. We used a modified FMEA technique and analyzed risks on the basis of probability and impact values and detectability score as shown in Fig. 3. The curve contains the risk 'R10' with the highest value of modified RPN and serves as a reference level of risk for assessing other risks. In contrast with the results of ranking scheme based on conventional method, R4 is not considered as an important risk. This scheme also treats risks as independent factors and therefore, the resulting ranking profile is not valid for the network setting considered in our simulation study.

RNELPM **Based Prioritization of Supply Chain Risks (Scheme 3).** This ranking scheme captured the interdependency between risks as depicted in Fig. 1. The relative importance of each risk is characterized by the associated value of *RNELPM* that can be assessed with reference to the curve containing the most critical risk as shown in

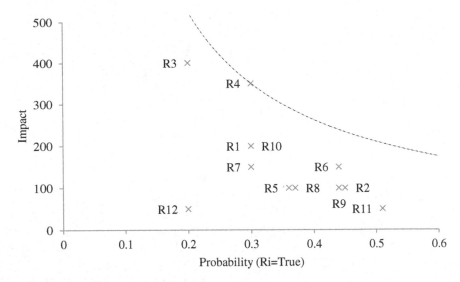

Fig. 2. Prioritization of supply chain risks based on conventional method

Fig. 4. As this scheme considers the attributes of probabilistic interdependency between risks, loss values and position of risks within an interconnected web of risks, the results reflect realistic nature of complex interactions between risks and risk sources. R11 appears to be the most significant risk taking into account its position in the network and strong relationship with connected nodes. However, this scheme assumes that all risks possess the same level of detectability. R3 and R12 are the least important risks because of combined effect of lower probability and loss values and weaker interdependency with the connected risks.

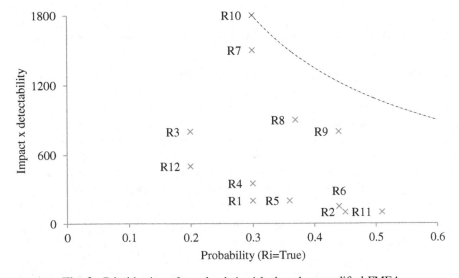

Fig. 3. Prioritization of supply chain risks based on modified FMEA

***RNEDIM* Based Prioritization of Supply Chain Risks (Scheme 4).** *RNEDIM* is an enhancement of the *RNELPM* as the *RNEDIM* based ranking scheme incorporates an important attribute of detectability in prioritizing risks. The risks are mapped on the basis of their detectability based propagation impact as shown in Fig. 5. Although the result of this scheme seems to be similar to that of the *RNELPM* based ranking scheme,

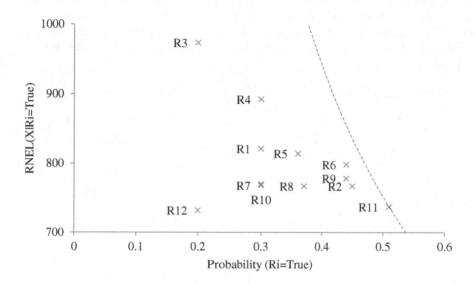

Fig. 4. *RNELPM* based prioritization of supply chain risks

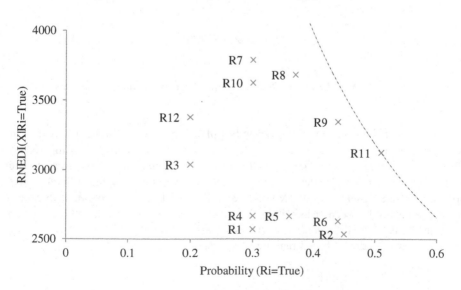

Fig. 5. *RNEDIM* based prioritization of supply chain risks

there are some obvious differences like in case of the ranking of R6, R7 and R8. Furthermore, it is important to understand that the unique array of loss and detectability values assumed in the study engendered such similar results.

RNELPM and RNEDIM Based Prioritization of Supply Chain Risks (Scheme 5). As the *RNEDIM* does not directly reflect the perceived risk exposure, it is important to relate the two ranking schemes based on *RNEDIM* and *RNELPM*. We normalized the two risk measures for each risk and prioritized risks through calculating the modulus of each vector associated with specific risk in the two-dimensional plane as shown in Fig. 6. This combined ranking scheme takes into consideration the detectability and contribution of each risk towards the network-wide propagation of loss. R11 (R12) is the most (least) important risk.

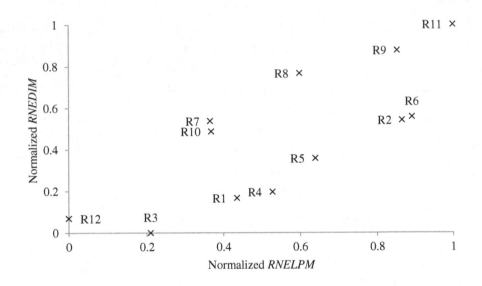

Fig. 6. *RNELPM* and *RNEDIM* based prioritization of supply chain risks

Comparison of Schemes for Prioritizing Supply Chain Risks. In order to realize the variability in results corresponding to the five ranking schemes, we compared the results as shown in Table 4. Generally, the results are similar for the last three schemes treating risks as interdependent factors whereas the results are quite different corresponding to the Schemes 1&2, 1&3 and 2&4 and therefore, it is important to capture interdependency between risks within the ranking framework. Furthermore, the variation in the detectability and/or loss values would yield quite different results even in case of interdependency based ranking schemes.

Table 4. Comparison of different prioritization schemes

Risk ID	Ranking of risks				
	Scheme 1	Scheme 2	Scheme 3	Scheme 4	Scheme 5
R1	4	10	8	10	10
R2	6	12	3	5	4
R3	2	5	11	12	11
R4	1	6	7	9	9
R5	10	8	5	8	6
R6	3	9	2	4	3
R7	6	2	10	6	7
R8	9	4	6	3	5
R9	8	3	4	2	2
R10	4	1	9	7	8
R11	11	11	1	1	1
R12	12	7	12	11	12

4.3   Managerial Implications

The proposed modelling approach can help supply chain managers prioritize supply chain risks taking into account the probabilistic interdependency between risks and loss and detectability associated with each risk. The approach is effective for assessing risks of complex supply chains as the risk network does not necessarily follow the process flow of the supply chain. The comparison of different ranking schemes can also help managers understand the limitations of each scheme and appreciate the importance of treating risks as interdependent factors. Causal mapping (qualitative modelling of BBNs) is beneficial to the managers in identifying important risks and understanding the dynamics between these risks.

5   Conclusion and Future Research

Limited studies have focused on capturing the interdependent interaction between risks within the literature of SCRM. Generally, risk registers are used for prioritizing risks that rely on the assumption of risks as independent factors. Detectability is an important concept in the FMEA that reflects the effectiveness of early warning system in detecting a risk before its complete realization. We adapted the concept of detectability from FMEA and integrated it within the theoretically grounded framework of BBNs comprising interconnected supply chain risks and associated loss values and demonstrated its application through an illustrative example. We also introduced two new risk measures for prioritizing supply chain risks and compared the results of our proposed ranking scheme with other schemes relying on independent or interdependent notion of risks. The results clearly revealed importance of considering detectability of risks in prioritizing supply chain risks as the risks having ineffective detection system must be given due consideration.

Our model is based on a number of assumptions. We have represented risks by binary states. In future, risks can be modelled as continuous variables and similarly, loss values can also be represented as continuous distributions. Our proposed method is a first step towards modelling detectability of risks within a framework of interdependent risks and risk sources. The proposed approach needs validation through conducting case studies.

References

1. Son, J.Y., Orchard, R.K.: Effectiveness of policies for mitigating supply disruptions. Int. J. Phys. Distrib. Logistics Manage. 43(8), 684–706 (2013)
2. Ackermann, F., Howick, S., Quigley, J., Walls, L., et al.: Systemic risk elicitation: using causal maps to engage stakeholders and build a comprehensive view of risks. Eur. J. Oper. Res. 238(1), 290–299 (2014)
3. Handfield, R., Blackhurst, J., Craighead, C.W., Elkins, D.: Introduction: a managerial framework for reducing the impact of disruptions to the supply chain (2011)
4. Standards, Risk Management: Principles and Guidelines (AS/NZS ISO 31000: 2009). Standards Australia, Sydney (2009)
5. Khan, O., Christopher, M., Burnes, B.: The impact of product design on supply chain risk: a case study. Int. J. Phys. Distrib. Logistics Manage. 38(5), 412–432 (2008)
6. Garvey, M.D., Carnovale, S., Yeniyurt, S.: An analytical framework for supply network risk propagation: a Bayesian network approach. Eur. J. Oper. Res. 243(2), 618–627 (2015)
7. Badurdeen, F., Shuaib, M., Wijekoon, K., Brown, A., et al.: Quantitative modeling and analysis of supply chain risks using Bayesian theory. J. Manuf. Technol. Manage. 25(5), 631–654 (2014)
8. Carbone, T.A., Tippett, D.D.: Project risk management using the project risk FMEA. Eng. Manage. J. 16(4), 28–35 (2004)
9. Gilchrist, W.: Modelling failure modes and effects analysis. Int. J. Qual. Reliab. Manage. 10 (5), 16–23 (1993)
10. Nepal, B., Yadav, O.P.: Bayesian belief network-based framework for sourcing risk analysis during supplier selection. Int. J. Prod. Res. 53(20), 6114–6135 (2015)
11. Jensen, F.V., Nielsen, T.D.: Bayesian Networks and Decision Graphs. Springer, New York (2007)
12. Qazi, A., Quigley, J., Dickson, A.: A novel framework for quantification of supply chain risks. In: 4th Student Conference on Operational Research, University of Nottingham, UK (2014)
13. Christopher, M., Lee, H.: Mitigating supply chain risk through improved confidence. Int. J. Phys. Distrib. Logistics Manage. 34(5), 388–396 (2004)
14. Sodhi, M.S., Tang, C.S.: Managing Supply Chain Risk. International Series in Operations Research and Mangement Science, vol. 172. Springer, New York (2012)
15. Bradley, J.R.: An improved method for managing catastrophic supply chain disruptions. Bus. Horiz. 57(4), 483–495 (2014)
16. Segismundo, A., Miguel, P.A.C.: Failure mode and effects analysis (FMEA) in the context of risk management in new product development. Int. J. Qual. Reliab. Manage. 25(9), 899–912 (2008)
17. Tuncel, G., Alpan, G.: Risk assessment and management for supply chain networks: a case study. Comput. Ind. 61(3), 250–259 (2010)

18. Lockamy, A., McCormack, K.: Analysing risks in supply networks to facilitate outsourcing decisions. Int. J. Prod. Res. **48**(2), 593–611 (2009)
19. Lockamy, A.: Benchmarking supplier risks using Bayesian networks. Benchmarking: Int. J. **18**(3), 409–427 (2011)
20. Lockamy, A., McCormack, K.: Modeling supplier risks using Bayesian networks. Industr. Manage. Data Syst. **112**(2), 313–333 (2012)
21. Lockamy, A.: Assessing disaster risks in supply chains. Industr. Manage. Data Syst. **114**(5), 755–777 (2014)
22. Dogan, I., Aydin, N.: Combining Bayesian networks and total cost of ownership method for supplier selection analysis. Comput. Industr. Eng. **61**(4), 1072–1085 (2011)
23. Qazi, A., Quigley, J., Dickson, A., Gaudenzi, B.: A new modelling approach of evaluating preventive and reactive strategies for mitigating supply chain risks. In: Corman, F., Voß, S., Negenborn, R.R. (eds.) ICCL 2015. LNCS, vol. 9335, pp. 569–585. Springer, Heidelberg (2015). doi:10.1007/978-3-319-24264-4_39
24. Qazi, A., Quigley, J., Dickson, A., Gaudenzi, B., et al.: Evaluation of control strategies for managing supply chain risks using Bayesian belief networks. In: International Conference on Industrial Engineering and Systems Management (2015)
25. Qazi, A., Quigley, J., Dickson, A., Gaudenzi, B., et al.: Selection of optimal redundancy strategies for a supply network. In: Kirsten, W., Blecker, T., Ringle, C.M. (eds.) Hamburg International Conference of Logistics (2015)
26. Onisko, A.: Medical diagnosis. In: Pourret, O., Naïm, P., Marcot, B. (eds.) Bayesian Networks: A Practical Guide to Applications, vol. 73, pp. 15–32. Wiley, West Sussex (2008)
27. Blodgett, J.G., Anderson, R.D.: A Bayesian network model of the consumer complaint process. J. Serv. Res. **2**(4), 321–338 (2000)
28. Qazi, A., Quigley, J., Dickson, A.: Supply chain risk management: systematic literature review and a conceptual framework for capturing interdependencies between risks. In: 5th International Conference on Industrial Engineering and Operations Management, Dubai (2015)
29. Sigurdsson, J.H., Walls, L.A., Quigley, J.L.: Bayesian belief nets for managing expert judgement and modelling reliability. Qual. Reliab. Eng. Int. **17**(3), 181–190 (2001)
30. GeNIe. The Decision Systems Laboratory, GeNIe and SMILE (2015). http://genie.sis.pitt.edu/about.html. Accessed 5 June 2015

E-Commerce Development Risk Evaluation Using MCDM Techniques

Salem Alharbi[✉] and Mohsen Naderpour

Faculty of Engineering and Information Technology, Global Big Data Technologies Centre,
University of Technology Sydney, Broadway, P.O. Box 123, Sydney, NSW 2007, Australia
Salem.A.Alharbi@student.uts.edu.au, Mohsen.Naderpour@uts.edu.au

Abstract. Electronic commerce (EC) development takes place in a complex and dynamic environment that includes high levels of risk and uncertainty. This paper proposes a new method for assessing the risks associated with EC development using multi-criteria decision-making techniques A model based on the analytic hierarchy process (AHP) and the technique for order of preference by similarity to ideal solution (TOPSIS) is proposed to assist EC project managers and decision makers in formalizing the types of thinking that are required in assessing the current risk environment of their EC development in a more systematic manner than previously. The solution includes the use of AHP for analyzing the problem structure and determining the weights of risk factors. The TOPSIS technique helps to obtain a final ranking among projects, and the results of an evaluation show the usefulness performance of the method.

Keywords: E-Commerce · Risk analysis · Multi-criteria decision making · AHP · TOPSIS

1 Introduction

Electronic commerce (EC) may be the most promising information technology application to emerge in recent years. EC addresses the needs of organizations, suppliers and customers to reduce costs while improving the quality of goods and services and increasing the speed of service delivery [1]. The current highly competitive business environment demands a high quality EC system; however, EC development is subject to various types of risk. Indeed, a task that is critical to the proper management of EC development is the assessment of risk. An important step in advancing our knowledge requires that we understand and address these risks.

The concept of risk became popular in economics during the 1920s. Since then, it has been successfully used in theories of decision making in economics, finance and decision science. Risk is defined as the "possibility of loss or injury" or "someone or something that creates or suggests a hazard". At present, there is no agreed upon universal definition of EC risk, but information security is a widely recognized aspect of EC risk [2].

Before conducting a risk analysis, the risk factors associated with EC development must be identified. Several empirical studies such as [3, 5] have focused on identifying

© Springer International Publishing Switzerland 2016
S. Liu et al. (Eds.): ICDSST 2016, LNBIP 250, pp. 88–99, 2016.
DOI: 10.1007/978-3-319-32877-5_7

the potential risk factors that threaten EC development. In the study by Wat et al. [3], a source-based approach to categorizing EC development risks is initially used, with technical, organizational and environmental risks as three primary source categories. Then the potential risks associated with EC development was identified, with 51 risk items based on a comprehensive literature review and interviews with EC practitioners. After an empirical study, 10 major dimensions of risks associated with EC development were proposed, namely (1) resources risk, (2) requirements risk, (3) vendor quality risk, (4) client–server security risk, (5) legal risk, (6) managerial risk, (7) outsourcing risk, (8) physical security risk, (9) cultural risk and (10) re-engineering risk. Ngai and Wat [2] used this classification and developed a web-based fuzzy decision support system for risk assessment, and Leung et al. [4] developed an integrated knowledge-based system that assists project managers in determining potential risk factors. In another study, Addison [5] used a Delphi technique to collect the opinions of experts and proposed 28 risks for EC projects. Meanwhile, Carney et al. [6] identified four categories comprising 21 risk areas and designed a risk evaluation tool. Cortellessa et al. [7] introduced a methodology which elaborates annotated Unified Modelling Language (UML) diagrams to estimate the performance failure probability and combined it with a failure severity estimate, which is obtained using functional failure analysis. This methodology has some limitations and is only suitable for the analysis of performance-based risk in the early phases of the software life cycle.

This paper develops a new risk evaluation method that can be used to effectively support EC project managers in conducting risk assessment in EC development. The idea relies upon the use of the analytic hierarchy process (AHP) and the technique for order of preference by similarity to ideal solution (TOPSIS), two popular multi-criteria decision-making (MCDM) techniques, in an innovative manner.

The paper is organized as follows. The background of this study is presented in Sect. 2. Section 3 shows our EC development risk evaluation method. The performance of the proposed method is illustrated in Sect. 4. Sections 5 and 6 summarize the discussion and conclusion.

2 Background

2.1 E-Commerce

Over the last two decades, the popularity of the Internet and network technology has increased rapidly. Consequently, EC has become a common activity in modern business operations [8]. The growth of EC activities within the last 20 years has attracted attention from academics as well as practitioners in various fields. For example, computer science researchers have shown interest in the technical and system sides of EC. Law academics are interested in the legal issues relative to EC. Business research focuses on the marketing and management issues of EC, while social research focuses on the influence of EC on human beings and society. From different perspectives, EC is variously referred to as follows [4]:

- Communication perspective: EC is the delivery of services/products, payments or information via telephone lines, computer networks or other means.
- Business process perspective: EC is the technology application for automating business workflows and transactions.
- Service perspective: EC is a tool for addressing the desire of consumers, management and firms to reduce service costs while increasing the quality of goods and service delivery speed.
- Online perspective: EC helps to sell and buy information and products on the Internet and through online services.

2.2   Multi-criteria Decision-Making (MCDM) Techniques

Description of MCDM Problems   MCDM is concerned with structuring and solving decision and planning problems involving multiple criteria. There are some common characteristics in MCDM problems, such as the presence of multiple non-commensurable and conflicting criteria, different units of measurement among the criteria and the presence of disparate alternatives. All the criteria in an MCDM problem can be classified into two categories. Criteria that are to be maximized are benefit criteria. In contrast, criteria that are to be minimized fall into the cost criteria category. A typical MCDM problem based on m alternatives $(A_1, A_2,..., A_m)$ and n criteria $(C_1, C_2, ..., C_n)$ can be presented as follows [9]:

$$\mathbf{X} = \left[x_{ij}\right]_{m*n}, \mathbf{W} = \left[w_j\right]_{n,} \tag{1}$$

where X is the decision matrix, x_{ij} is the performance of the ith alternative with respect to the jth criterion, W is the weight vector and w_j is the weight of the jth criterion. The original decision matrix X is typically incomparable, because different criteria are expressed using different units of measure. Therefore, data should be transformed into comparable values using a normalization procedure. The weight vector W has a large effect on the ranking results of alternatives. It is usually fixed using an expert's assignment or the AHP method [9].

AHP is a structured technique for organizing and analyzing complex decisions based on mathematics and psychology. It was developed by Thomas L. Saaty in the 1970s and has been extensively studied and refined since then. AHP relies on the judgment of an expert to establish a priority scale and uses pairwise comparisons. It is a highly popular MCDM method, and it has certain advantages and disadvantages [10]. One of the biggest advantages of AHP is that it is easy to use. As this method utilizes pairwise comparisons, it allows the coefficients to be weighted and the alternatives to be easily compared. Due to its hierarchical structure, it is scalable and can be adjusted in size to accommodate decision-making problems. AHP requires data for pairwise comparison, but it is not as data-intensive as other MCDM methods such as multi-attribute utility theory. One of the common limitations of AHP is that it experiences problems related to the interdependence between the criteria and alternatives [11]. The AHP approach involves pairwise comparison, and thus it is vulnerable to inconsistencies in ranking criteria and judgment. AHP evaluates each instrument in comparison with the other instruments and does not grade any instrument in isolation. Therefore, the approach is unable to properly identify the weaknesses and strengths associated with each instrument [12]. AHP is susceptible to rank reversal in its general form, and this is another major drawback. As rankings are used for comparison, the addition of alternatives at the end of the process can result in the final rankings reversing or flipping. AHP has been used extensively for resource management, performance-type problems, corporate strategy and policy, political strategy and public policy and planning. Problems related to resource management have a limited number of alternatives, which minimizes the disadvantage of rank reversal.

AHP breaks the decision-making process into steps so that decisions can be made in an organized way with defined priorities. The steps for decision making in AHP are as follows [10]:

Step 1. Define the problem: In this step, the problem is defined, and the goal is determined.
Step 2. Represent the problem graphically: Information is arranged into a hierarchical structure. The decision goal is stated, the criteria are defined and the alternatives are identified.
Step 3. Develop a judgment preference: This is done to measure the preference for alternatives against the criteria. A pairwise comparison is made on a scale of one to nine to rate the relative preference.
Step 4. Relative weight calculation: The relative weights of the alternatives and criteria are determined through calculations. The eigenvalue technique is used to calculate the priority vector or relative weight, where P is the priority vector.
Step 5. Synthesis: By aggregating the weights in the results vertically, the contribution of every alternative to the overall goal is computed. An overall ranking of the alternatives is obtained through combining the priority vectors of all the criteria.
Step 6. Consistency: The consistency ratio is used to measure the accuracy of the decision.

TOPSIS. Among various methods that have been established to solve real-world MCDM problems, TOPSIS works satisfactorily in various situations and can be applied to diverse areas. Hwang and Yoon were the first to propose this technique, which can be used for selecting and evaluating the best alternative for a problem [13]. It has become a well-known classical MCDM method and has gained ample interest from practitioners and researchers around the world.

TOPSIS is a ranking method that is easy to understand and apply. It aims to select alternatives that are farthest from the negative ideal solution and closest to the positive ideal solution. The advantage of choosing the positive ideal solution is that it minimizes cost criteria and maximizes benefit criteria, whereas the negative ideal solution minimizes the benefit criteria and maximizes the cost criteria. The use of TOPSIS helps in arranging alternatives cardinally, making maximum use of attribute information; therefore, it does not require independent attribute preference [14]. The application of TOPSIS requires attribute values to be monotonically decreasing or increasing and numeric and to have commensurable units. Figure 1 shows the procedure for implementing TOPSIS [15]. First, the initial decision matrix is formed, and then the decision matrix is normalized. In the second step, the weighted normalized decision matrix is constructed, and this matrix is then used in the third step to determine the negative and positive ideal solutions. In the fourth step, the measures of separation for every alternative are calculated. In the fifth and final step, the coefficient of relative closeness is computed. Then, according to the value of the coefficient of closeness, alternatives are ranked or arranged in increasing order. The TOPSIS process is performed as follows [9]:

Step 1. Normalize the decision matrix:

$$r_{ij} = \frac{x_{ij}}{\sqrt{\sum_{i=1}^{m} x_{ij}^2}}, i = 1, \ldots, m; j = 1, \ldots, n, \tag{2}$$

where r_{ij} denotes the normalized value of the jth criterion for the ith alternative A_i.

Step 2. Calculate the weighted normalized decision matrix:

$$v_{ij} = r_{ij} * w_j, i = 1, \ldots, m; j = 1, \ldots, n, \tag{3}$$

where w_j is the weight of the jth criterion or attribute.

Step 3. Determine the positive ideal and negative ideal solutions:

$$A^+ = \left(v_1^+, \ldots, v_n^+\right) \tag{4}$$

$$A^- = \left(v_1^-, \ldots, v_n^-\right), \tag{5}$$

where A^+ denotes the positive ideal solution, and A^- denotes the negative ideal solution. If the jth criterion is a beneficial criterion, then $v_j^+ = \max\{v_{ij}, i = 1, \ldots, m\}$ and $v_j^- = \min\{v_{ij}, i = 1, \ldots, m\}$. In contrast, if the jth criterion is a cost criterion, then $v_j^+ = \min\{v_{ij}, i = 1, \ldots, m\}$ and $v_j^- = \max\{v_{ij}, i = 1, \ldots, m\}$.

Step 4. Calculate the distances from each alternative to a positive ideal solution and a negative ideal solution:

$$D_i^+ = \sqrt{\sum_{j=1}^{n} (v_{ij} - v_j^+)^2}, i = 1, \ldots, m \tag{6}$$

$$D_i^- = \sqrt{\sum_{j=1}^{n} (v_{ij} - v_j^-)^2}, i = 1, \ldots, m, \tag{7}$$

where D_i^+ denotes the distance between the ith alternative and the positive ideal solution, and D_i^- denotes the distance between the ith alternative and the negative ideal solution.

Step 5. Calculate the relative closeness to the ideal solution:

$$C_i = \frac{D_i^-}{D_i^+ + D_i^-}, i = 1, \ldots, m. \tag{8}$$

Step 6. Rank the alternatives, sorting by the value C_i in decreasing order.

3 The E-Commerce Development Risk Evaluation Method

As discussed earlier, EC offers many opportunities for business, and EC projects have many advantages. Even so, EC development is subject to various risks, and it is essential to manage these risks to avoid problems. The assessment of risk is thus essential to the proper management of EC projects. This section develops a risk evaluation method for EC development. Firstly, the risk factors are introduced in Sect. 3.1. Secondly, the method is presented in Sect. 3.2.

3.1 Risk Factors

Several empirical studies have been carried out examining the risk factors that threaten EC project development. More than 50 risk factors have been identified in the literature. However, considering all of them is not practical in any situations. In this paper, 12 important risk factors, based on the knowledge of our expert, who has seven years of professional experience in EC project management, have been considered. The factors are categorized in three groups as shown in Fig. 1, including technical, organizational and environmental risks. In addition, 12 major risk factors are determined as shown in Table 1. The table shows the risk categories and potential risks associated with EC development in each category.

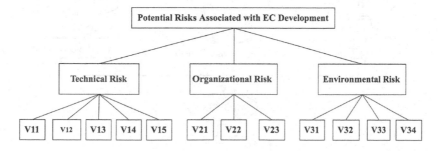

Fig. 1. The hierarchy structure of EC development risks.

Table 1. Potential risks associated with EC development.

Category	Variable	Potential risk
Technical risk	V11	Project complexity
	V12	Software or hardware problem-caused system failure
	V13	Poor design, code or maintenance procedure
	V14	Wrong functions and properties development
	V15	Continuous change of system requirements
Organizational risk	V21	Wrong schedule estimation
	V22	Project over budget
	V23	Lack of expertise and experience in EC
Environmental risk	V31	Lack of international legal standards
	V32	Difficult to change outsourcing decision/vendor
	V33	Loss of data control
	V34	Different users with different cultures

3.2 Method

The risk evaluation method, shown in Fig. 2, consists of two parts: (1) determining the weights of the criteria and (2) evaluating the alternatives. The goal is to select the project that has the least amount of associated risk. The risk factors identified in the previous section are considered as criteria. Therefore, the pairwise comparison of AHP is used

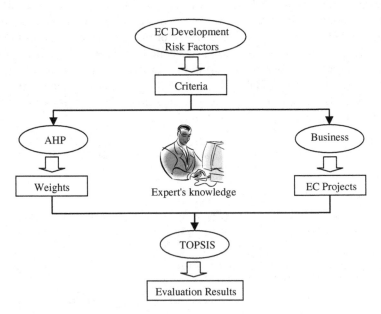

Fig. 2. The EC risk evaluation method. In order to determine the weights of the risk factors, the AHP methodology is applied. Once the weights have been obtained, the evaluation of the EC projects can be carried out using the TOPSIS methodology.

to calculate the weights of the criteria and sub-criteria. Once the weights have been obtained, the evaluation of the alternatives is carried out using the TOPSIS methodology.

4   Application

EC projects generally have a high IT content (around 20–40 %), but they differ from IT projects in various ways. They are relatively low in cost and are characterized by high business impact. Moreover, EC projects have complex architecture and a high need for fault tolerance and scalability. They require a high rate of continuous evolution, which is driven by competitive actions, technology changes and business model innovations. EC projects are widely used to promote retailing. In the following, three EC projects of a private company are introduced. The projects include an online shopping mall, an online auction and an online bookstore.

4.1   Projects of Interest

- Online shopping mall: The company would like to develop an online multi-vendor shopping mall that will allow customers to purchase products from participating online stores. The online mall will allow prospective customers to select products from a range of categories/sub-categories and give them the opportunity to compare products from various vendors and easily view special offers before adding their choices to a shopping cart and participating in a secure transaction. Prospective vendors should be able to easily pay an annual registration fee, create their own online stores and take control of their stores and add process orders and products from customers.
- Online auction: The online auction is a group that would be established for the purpose of auctions. If an individual wants to sell something via an auction, they would post on that website. The business is simply a selling of products, which gives the project user the ability to bid on particular services or products. The primary objective of the e-auction process will be to obtain the highest price and best value. It is not possible to achieve the best value outcomes when the focus remains on price. There are two categories of persons in this project: customers and vendors. Both have their own registration forms. Vendors can sell their products on this website, and customers can purchase them. Each product will be awarded to the customer who places the highest bid on the product.
- Online bookstore: This project's main objective is to create an online bookstore. This bookstore will give users a chance to search for and purchase books online by author, title and subject. A tabular format is used to display selected books, and the online bookstore will facilitate the user in making online purchases using a credit card. Customer using this website can easily purchase books online and do not need to visit a bookstore and waste time.

4.2   Weighting Criteria Using AHP

Based on the proposed method, the ratings are obtained from an expert who has seven years of professional experience in EC project management and is familiar with legal services. The matrices are formed and priorities are synthesized using the AHP methodology. Table 2 shows the expert's knowledge in a pairwise comparison of categories. Then, the weights for the risk categories are calculated by AHP.

Table 2. Pairwise comparison matrix for risk categories.

	Technical risk	Organizational risk	Environmental risk	Geometric mean	Normalized
Technical risk	1.00	0.25	0.17	0.347	0.089
Organizational risk	4.00	1.00	0.50	1.260	0.323
Environmental risk	6.00	2.00	1.00	2.289	0.588
Sum				3.896	1.000

The results in Table 2 are summarized in Fig. 3. As can be seen, environmental risk is the most important risk category. It is followed by organizational risk and technical risk.

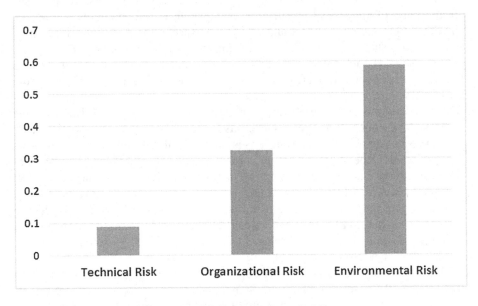

Fig. 3. The calculated weights for each risk category.

The complete priority weighting and ranking of the risks are shown in Table 3. The results show that lack of international legal standards is the most significant risk (0.305), and project complexity is the least significant risk. Overall, environmental risks have the highest global priority weighting, as mentioned previously.

Table 3. The weight of each sub-criteria.

Sub-category	Technical Risk
Project complexity	0.003
Software or hardware problem-caused system failure	0.011
Poor design, code or maintenance procedure	0.024
Wrong functions and properties development	0.006
Continuous change of system requirements	0.045
Sub-category	Organizational Risk
Wrong schedule estimation	0.030
Project over budget	0.070
Lack of expertise and experience in EC	0.223
Sub-category	Environmental Risk
Lack of international legal standards	0.305
Difficult to change outsourcing decision/vendor	0.173
Loss of data control	0.072
Different users with different cultures	0.038
SUM	1.000

Table 4. The expert's opinion for EC projects under each risk factor.

Technical risk					
Sub-category	V11	V12	V13	V14	V15
Global weight (criteria*sub-criteria)	0.003	0.011	0.024	0.006	0.045
Project 1: Online shopping mall	1	5	6	5	7
Project 2: Online auction	4	3	5	3	4
Project 3: Online bookstore	2	5	2	1	1
Organizational risk					
Sub-category	V21		V22		V23
Global weight (criteria*sub-criteria)	0.03		0.07		0.223
Project 1: Online shopping mall	4		6		6
Project 2: Online auction	6		2		6
Project 3: Online bookstore	3		1		7
Environmental risk					
Sub-category	V31	V32	V33		V34
Global weight (criteria*sub-criteria)	0.305	0.173	0.072		0.038
Project 1: Online shopping mall	1	5	1		4
Project 2: Online auction	7	1	1		5
Project 3: Online bookstore	6	3	2		5

4.3   Ranking EC Projects Using TOPSIS

Table 4 shows the matrix based on the expert's score for each EC project in each risk factor. We start by calculating the weighted normalized matrix to find the ideal solutions.

Several matrices, including the square roots of all sub-criteria, the weighted normalized distance to positive ideal solution and distance to negative ideal solution are omitted for space consideration. Table 5 shows the relative closeness to the ideal solution for each project. As can be seen, the online shopping mall is considered to be the best solution for this case.

Table 5.   The relative closeness to the ideal solution.

Alternative	TOPSIS scores
Project 1: Online shopping mall	0.599705
Project 2: Online auction	0.397074
Project 3: Online bookstore	0.345942

5   Discussion

In this study, several important risk factors threatening EC projects have been identified. In the proposed method, AHP was used for finding the priority weight of the risks involved, but it was not used for ranking the alternatives, because it would require a large number of matrices. Instead, TOPSIS, which avoids the complexity related to AHP, was used for ranking the alternative EC projects. TOPSIS is also simpler and faster than AHP in application and makes it possible to treat the qualitative variables quantitatively. This helps the decision making become simpler by providing positive and negative ideal solutions, allowing the decision maker to easily determine which alternative to choose. Moreover, TOPSIS is not limited to the number of criteria that are being applied. A case study containing three EC projects for a particular company was considered for risk evaluation.

6   Conclusion and Future Work

The paper developed an approach for selecting EC projects with minimum risks. AHP methodology was used to calculate the weights of the different risk factors contributing to EC projects. TOPSIS methodology was applied to rank the alternatives, thus allowing the selection of the best alternative project. The results show the significance of AHP and TOPSIS in solving MCDM problems in EC projects when managers are concerned about choosing the alternative with the lowest risks and maximum benefits.

There are various future directions for this research. Fuzzy logic can be used for assessing the risks of EC projects. Fuzzy logic can allow mangers to use linguistic variables, and this can help them to perform better risk assessments such as conducting failure modes and effects analysis (FMEA). Fuzzy logic can be used along with AHP and TOPSIS to solve complex risk problems in EC projects. We will also build a prototype based on the proposed method, and the results will be compared with other EC risk assessment methodologies as well.

References

1. Kalakota, R., Whinston, A.B.: Electronic Commerce: a Manager's Guide. Addison-Wesley Professional, Reading (1997)
2. Ngai, E.W.T., Wat, F.K.T.: Fuzzy decision support system for risk analysis in e-commerce development. Decis. Support Syst. **40**, 235–255 (2005)
3. Wat, F., Ngai, E.W., Cheng, T.E.: Potential risks to e-commerce development using exploratory factor analysis. Int. J. Serv. Technol. Manag. **6**, 55–71 (2005)
4. Leung, H.M., Rao Tummala, V.M., Chuah, K.B.: A knowledge-based system for identifying potential project risks. Omega **26**, 623–638 (1998)
5. Addison, T.: E-commerce project development risks: evidence from a Delphi survey. Int. J. Inf. Manag. **23**, 25–40 (2003)
6. Carney, D.J., Morris, E.J., Place, P.R.: Identifying commercial off-the-shelf (COTS) product risks: the COTS usage risk evaluation. Carnegie-Mellon Univ Pittsburgh (2003)
7. Cortellessa, V., Goseva-Popstojanova, K., Appukkutty, K., Guedem, A.R., Hassan, A., Elnaggar, R., Abdelmoez, W., Ammar, H.H.: Model-based performance risk analysis. IEEE Trans. Softw. Eng. **31**, 3–20 (2005)
8. Wang, C.-C., Chen, C.-C.: Electronic commerce research in latest decade: a literature review. Int. J. Electron. Commerce Stud. **1**, 1–14 (2010)
9. Wang, P., Zhu, Z., Wang, Y.: A novel hybrid MCDM model combining the SAW, TOPSIS and GRA methods based on experimental design. Inf. Sci. **345**, 27–45 (2016)
10. Saaty, T.L.: Decision making with the analytic hierarchy process. Int. J. Serv. Sci. **1**, 83–98 (2008)
11. Saaty, T.L.: Fundamentals of Decision Making and Priority Theory with the Analytic Hierarchy Process. Rws Publications, Pittsburgh (2000)
12. Konidari, P., Mavrakis, D.: A multi-criteria evaluation method for climate change mitigation policy instruments. Energy Policy **35**, 6235–6257 (2007)
13. Behzadian, M., Otaghsara, S.K., Yazdani, M., Ignatius, J.: A state-of-the-art survey of TOPSIS applications. Expert Syst. Appl. **39**, 13051–13069 (2012)
14. Hwang, C.-L., Yoon, K.: Multiple Attribute Decision Making: Methods and Applications a State-of-the-Art Survey. Springer Science & Business Media, Heidelberg (2012)
15. Wu, F.-Y., Chuang, C.-C.: The optimal relationship between buyer and seller obtained using TOPSIS method. J. Adv. Manag. Sci. **1**, 133–135 (2013)

Scaling Issues in MCDM Portfolio Analysis with Additive Aggregation

Carolina Lino Martins, Jonatas Araujo de Almeida[✉],
Mirian Batista de Oliveira Bortoluzzi, and Adiel Teixeira de Almeida

Production Engineering Department, Federal University of Pernambuco,
Av. Acadêmico, Hélio Ramos, s/n, Cidade Universitária, Recife, PE, Brazil
carol_tcch@hotmail.com, jonatasaa@yahoo.com.br,
mirianbortoluzzi@gmail.com, almeidaatd@gmail.com

Abstract. This paper discusses a typically scaling issue, which can arise in the context of multicriteria (MCDM) portfolio analysis: the portfolio size effect. By analyzing previous application this issue may happen by the impact of an additive aggregation for the standard portfolio construction model. Thus, it has been shown that the scaling issue may arise even when baseline correction procedures are adopted and this paper suggests that additionally to the baseline adjustment, a ratio scale correction may be necessary, depending on the combination of values and constraints considered by the problem.

Keywords: Project portfolio · Portfolio scaling issue · Portfolio size effect · Baseline in portfolio

1 Introduction

Project portfolio selection problems involve the selection of a set of projects, considering different aspects and taking into account some constraints given by the context, to be undertaken by a company that seeks to optimize the general portfolio value [2].

Optimal portfolios can be readily determined with multi-attribute decision methods that use mathematical programing techniques to construct them [7]. For the decision maker, this simplifies the multi-attribute evaluation task, which has to be performed only on individual items. The number of items is much smaller than the total number of (feasible and efficient) portfolios, and the consequences of individual items might be easier to evaluate. This makes this approach particularly suitable for Decision Support Systems, in the case of portfolio problems [7].

There are plenty of methods from multi-attribute decision making to evaluate the items from a portfolio, such as outranking methods like PROMETHEE [13], Data Envelopment Analysis [5], or additive utility functions [8], that will be the focus of this paper.

In such cases, an additive value function aggregates the projects' attribute-specific performances into an overall project value, and the portfolio value is the sum of those projects' overall values that are included in the portfolio [9]. This method commonly

S. Liu et al. (Eds.): ICDSST 2016, LNBIP 250, pp. 100–110, 2016.
DOI: 10.1007/978-3-319-32877-5_8

attributes the value of 0, a baseline measurement, to the worst item of a specific criteria analyzed in the portfolio [12].

Clemen and Smith [4] argue that this implicit baseline is often inappropriate and may lead to incorrect recommendations when practitioners assume that not doing a project results in the worst possible score on all attributes. Hence, the authors make some assumption about how to evaluate not doing a project correctly. These baseline corrections have also been examined by Liesiö and Punkka [9].

Even considering the importance of settling an appropriate baseline for the problem, it is extremely necessary to be aware that additive utility functions approach imposes certain requirements on the measurement scales used for the items in a portfolio, which are frequently ignored in existing literature [7] and have considerable impact on the results when they are not taken into account. Also, there are limited computational tools to analyze and specify these issues.

Thus, the purpose of this paper is to make an application adapted from the problem proposed by Clemen and Smith [4] and to show that the major problem it is not to determine a baseline, but perhaps, the scaling issue that exists in additive multicriteria portfolio analysis. Consequently, this paper is not about the baseline specifically and the baseline does not exist in every problem, as explained by [12] and it is not present in this application either.

The numerical application of this research was assessed using a computational tool [3], which makes a MCDM additive portfolio analysis via web for linear intra-criteria value function with sensitivity analysis, using Monte Carlo simulation.

Also, this work is an extension from previous studies mentioned in the text, such as: [1, 6, 7, 13], which are related to MCDM portfolio analysis in different perspectives and give support for the understanding of the problem presented here.

The paper is structured as follows. Section 2 makes a literature review and some considerations on the baseline problem. Section 3 discusses different scaling issues aspects. In Sect. 4, a numerical application is presented, and, finally, Sect. 5 concludes the paper.

2   Literature Review and Considerations on the Baseline Problem

Different papers have discussed about the baseline problem that arose from Clemen and Smith [4] work, who noted that a model of the form

$$\sum_{i=1}^{n} z_i v(A_i)$$ (1)

where z_i is a binary variable indicating whether item A_i is included in the portfolio ($z_i = 1$ if it is included and $z_i = 0$ if it is not), and $v(A_i)$ is the value of item A_i obtained from the multi-attribute evaluation, implies that the outcome of not performing a project has a utility of zero [4].

In the usual scaling of marginal utility functions, this would mean it is identical to the worst possible outcome. This condition is clearly not always fulfilled, in particular if some attributes refer to negative effects of the items [7]. The authors pointed out that the utility scale should be chosen in a way that zero utility is assigned to the outcome of not doing a project, rather than the worst possible outcome, which implies that some projects have negative marginal utility values indicating that the project worsens outcomes in some attributes [7].

Morton [12] underscore the criticism of the value function of not doing a project by showing it can lead to a rank reversal and provide a measurement theoretic account of the problem, showing that the problem arises from using evaluating projects on an interval scale whereas to guard against such rank reversals, suggesting that the benefits must be measured on at least a ratio scale.

Liesiö and Punkka [9] presented a baseline value specification technique that admit incomplete preference statements and make it possible to model problems where the decision maker would prefer to implement a project with the least preferred performance level in each attribute. They also show how these results can be used to analyze how sensitive project and portfolio decision recommendations are to variations in the baseline value and provide project decision recommendations in a situation where only incomplete information about the baseline value is available.

de Almeida et al. [7] discuss the effects of different utility scales on the results of multi-attribute portfolio problems. They analyze three effects: the portfolio size effect, the baseline effect and consistency across different aggregation sequences. They also show that these three effects have similar causes related to the use of an interval utility scale, which allows for additive transformation of utilities.

In [6] the problem noted by Mavrotas et al. [10] is analyzed. This problem related to PROMETHEE V method for multi-attribute analysis, which fails to include an item in a portfolio if it has a negative net flow with respect to other items. They pointed out that the model formulated by Mavrotas et al. [10] introduces a bias in favor of large portfolios because the PROMETHEE V method is sensitive to scale transformations, so de Almeida and Vetschera [6] propose to use the c-optimal portfolio concept in order to overcome this issue, which has been applied in [1].

Vetschera and Almeida [13] also explore a new formulation of the PROMETHEE V method and develop several alternative approaches based on the concepts of boundary portfolios and c-optimal portfolios.

Even considering the importance given by the baseline problem, it is worthwhile to note that the additive utility functions approach imposes certain requirements on the measurement scales used for the items in a portfolio that should not be ignored, given the different results that can arise from the portfolio size effect, as it will be shown in Sect. 3 and in the numerical adaptation of the paper.

Therefore, this paper suggests that the major problem it is not to define the baseline, but perhaps, to make an adequate scale transformation that it is appropriate for additive multicriteria portfolio analysis. Nevertheless, if there is a baseline problem in multiattribute portfolio analysis, then it is important not to forget to verify scale requirements.

In addition, the scale transformation pointed out by this paper was applied to the same problem proposed by Clemen and Smith [4] but did not present the portfolio size effect, once the combination of values was not favorable for the case. Thus, a number of instances was generated until find the one showed in Sect. 4, which kept the value structure considered by [4] and taking into account factual data.

Consequently, it is possible to infer that the portfolio size effect does not happen for all the cases and they will depend on the combination of values and constraints considered by the problem analyzed. Additionally, it is always important to examine the existence of the scale problem and, if it does happen, then one should make the necessary changes to adequate the case.

3   Scaling Issues Aspects

A model of the form (1), previously presented, is not invariant to a linear transformation of the value functions [7]. Though, a transformation of scores is sometimes needed with the aim of avoiding to exclude portfolios with a negative net flow [10]. This can easily be shown, as already pointed out in [6], for the PROMETHEE V and can also be applied for the additive model, through replacing the value function $v(.)$ by a function $w(.) = av(.) + b$:

$$max \sum_{i=1}^{n} z_i w(A_i) = \sum_{i=1}^{n} z_i (av(A_i) + b) = a \sum_{i=1}^{n} z_i v(A_i) + bc \qquad (2)$$

where $c = \sum z_i$ is the number of items contained in the portfolio. Depending on the sign of b, a linear transformation of the original value function will thus lead to a different objective function which favors either large portfolios (for $b > 0$) or small portfolios (for $b < 0$). de Almeida et al. [7] denoted this effect as the *portfolio size effect*.

In [13] the concept of c-optimal portfolios is proposed to overcome the portfolio size effect. By adding the constraint $c = \sum z_i$, problem (1) can be solved for portfolios of a given size c. By varying c, different portfolios are obtained, which then can be compared to each other at the portfolio level using any multi-attribute decision method. Nevertheless, for additive models, the portfolio size effect do not exist if $v(A_i)$ is measured on a ratio scale, which has a fixed zero point and to solve the baseline problem, should be identical to the outcome of not including an item in the portfolio [7].

To obtain equivalent evaluations of alternatives, the weights must be rescaled as:

$$\overline{q_j} = k_j \cdot (\overline{x_j}/\overline{x_j} - x_j) \qquad (3)$$

where $x_j = \min_i x_{ij}$ is the worst and $\overline{x_j} = \max_i x_{ij}$ is the best outcome in attribute j. k_j represents the weights used in the original model using an interval scale and q_j the weights to be used for a ratio scale.

To see detailed information about this transformation or other topics on the subject, see [7].

Following numerical application shows that this issue may arise, even if baseline corrections are introduced, as already pointed out by Morton [11].

4 Numerical Application with Baseline Correction

The ideas presented in this paper can be seen in an application, which is an adaptation of the one given by Clemen and Smith [4].

It was considered a portfolio of Information Technology projects, which are evaluated according to three criteria: financial contribution, risk and fit, as shown in Table 1. There are eight projects (A – H) to consider. The attribute financial contribution is measured in dollars. Risk reflects the probability that the project will lead to a marketable product [4] and is classified as: safe (least risky projects), probable (intermediate level) and uncertain (riskiest projects). Fit could be a rough measure of an incremental revenue that the project might generate [4] and is scored on a scale from 1 (worst) to 5 (best). The last column lists the necessary days required for each project, in total there is a limit of 2500 person days. The weights presented in the last line are related to an elicitation procedure based on an interval scale and the scores are already normalized in Table 2.

Table 1. Data for example.

Project	Financial contribution	Risk	Fit	Days required
A	392913	Uncertain	5	700
B	227503	Safe	2	400
C	155012	Probable .	3	600
D	136712	Uncertain	3	250
E	441713	Uncertain	5	300
F	382780	Probable	3	350
G	202678	Probable	5	600
H	189295	Safe	1	800
Weights	**0,25**	**0,25**	**0,5**	

Table 2. Data for example with normalized scores for each attribute

Project	Financial contribution	Risk	Fit	Value score	Days required	Go?
A	0,84	0	1	0,710	700	1
B	0,30	1	0,25	0,449	400	1
C	0,06	0,5	0,5	0,390	600	0
D	0,00	0	0,5	0,250	250	0
E	1,00	0	1	0,750	300	1
F	0,81	0,5	0,5	0,577	350	1
G	0,22	0,5	1	0,679	600	1
H	0,17	1	0	0,293	800	0
Weights	**0,25**	**0,25**	**0,5**			

Applying the additive model, the results indicate a portfolio with projects A, B, E, F and G, for the interval scale. This solution uses a total of 2350 days, 150 less than the total available.

On the other hand, using a ratio scale with the appropriate new set of weights, leads to another portfolio, in which A is replaced by C and D, and this is the correct solution for the problem, based on a multicriteria portfolio analysis. In this case, the interval scale favors a portfolio with size $c = 5$, while the ratio scale indicates a portfolio with $c = 6$. Projects C and D require 150 additional days, available in the limit constraint, compared to project A. Thus, the solution uses a total of 2500 days. Besides, these two projects give additional outcomes for risk and fit criteria when compared to project A. The new results are shown in Tables 3 and 4 shows the comparison between alternatives A, C and D.

Table 3. Results for a ratio scale

Project	Financial contribution	Risk	Fit	Value score	Days required	Go?
A	0,89	0	1	0,766	700	0
B	0,52	1	0,4	0,555	400	1
C	0,35	0,5	0,6	0,507	600	1
D	0,31	0	0,6	0,394	250	1
E	1,00	0	1	0,798	300	1
F	0,87	0,5	0,6	0,658	350	1
G	0,46	0,5	1	0,741	600	1
H	0,43	1	0,2	0,429	800	0
Weights	**0,29**	**0,20**	**0,51**			

Table 4. Comparison between alternatives

Project	Financial contribution	Risk	Fit	Total value score
A	0,89	0	1	0,766
C + D	0,66	0,5	1,2	0,901

The interval scale favors portfolios with fewer alternatives, decreasing artificially the actual value of larger portfolios. This application is realistic and similar to any kind of portfolio related to real application, particularly in the domain of information systems and DSS Web [7].

These results from the numerical application of this paper were easily assessed using a computational tool [3], which makes a MCDM additive portfolio analysis via web for linear intra-criteria value function with sensitivity analysis, using Monte Carlo simulation.

The program is divided into eight parts: the main page of the system; the input data part; a page that shows the consequences matrix, constraints, parameters and weights from the problem; an option to transform or not the input weights, depending on if they were obtained in an elicitation context of ratio scale or interval scale; an option page of not transforming the input weights or an option page when transforming the input weights; the sensitivity analysis part; and, finally, the results from the sensitivity analysis.

A few parts from the program are shown next.

In the input data page, the user can choose a problem, put its name and make a description (1 from Fig. 1) of the problem. It is possible to choose a file and import a worksheet (2 from Fig. 1), according to the worksheet model from the system, and there is the description from the criteria and alternatives (3 from Fig. 1), that shows up automatically when the worksheet is imported.

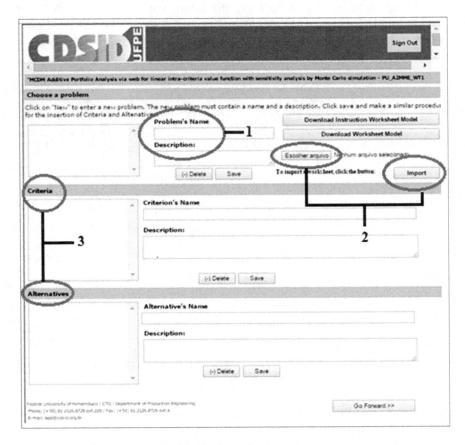

Fig. 1. Input data page.

The option page of transforming the weights shows a comparison between the weights with interval scale (1 from Fig. 2) and the weights obtained with a ratio scale (2 from Fig. 2), both of them for each criteria. (3 from Fig. 2), (4 from Fig. 2), (5 from Fig. 2) are the results vector for each alternative, the alternatives portfolio value and the overall portfolio value, respectively.

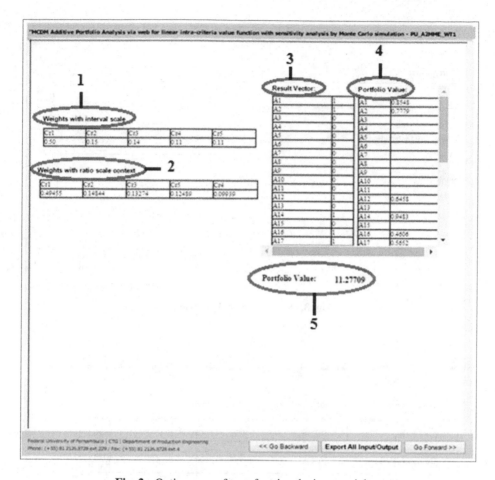

Fig. 2. Option page of transforming the input weights.

Figure 3 shows the sensitivity analysis page. Here, the user can choose the parameters, weights (1 from Fig. 3) and consequences matrix (2 from Fig. 3), the probability distribution (3 from Fig. 3), uniform or triangular, the number of cases (4 from Fig. 3) and the range (5 from Fig. 3).

Figure 4 shows the page of results from the sensitivity analysis, in which there are the cases with non-standard portfolios (1 from Fig. 4), the cases with standard portfolios (2 from Fig. 4), the total cases (3 from Fig. 4), the number of non-standard portfolios (4 from Fig. 4), the parameters (5 from Fig. 4), the range (6 from Fig. 4) and, finally, the chosen probability distribution (7 from Fig. 4).

The system also offers the possibility of exporting all input/output from the problem to a Microsoft Excel Worksheet, if the user needs.

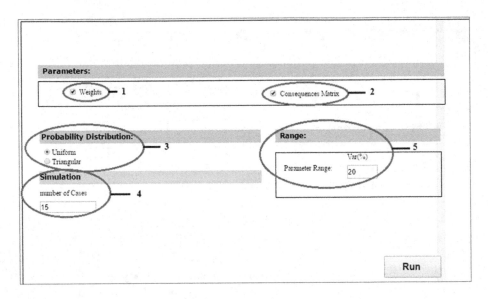

Fig. 3. Sensitivity analysis page

A24	0				0		Times
A25	0				0		Times
A26	1				0		Times
A27	1				0		Times
A28	0				0		Times
A29	1				6		Times
A30	0				3		Times
A31	1				0		Times
A32	1				1		Times
A33	0				0		Times
A34	1				1		Times
A35	1				0		Times
A36	1				0		Times
A37	1				0		Times
		1	Cases with non-standard portfolios:	10	66.67%		
		2	Cases with standard portfolios:	5	33.33%		
		3	Total Cases:	15	100%		
		4	Non-standard portfolios:	8			
		5	Parameter:	All Parameters			
		6	Range:	20%			
		7	PDF:	Uniform			

"MCDM Additive Portfolio Analysis via web for linear intra-criteria value function with sensitivity analysis by Monte Carlo simulation - PU_A2MME_WT1

Sensitivity Analysis   Portfolios Found   Export All Input/Outputs

Fig. 4. Page of results from the sensitivity analysis

5 Conclusions

In this paper, it was discussed a typically scaling issue which can arise in the context of multi-attribute portfolio problems: the portfolio size effect that additive changes in the utility of items create a bias in portfolio evaluation which depends on the number of items in the portfolio. It was shown, by a numerical application adapted from the problem proposed by Clemen and Smith [4], that this issue is caused by the impact of an additive utility transformation on the standard portfolio construction model.

Even considering the importance given by the baseline problem in different papers, this paper pointed out that it is valuable to note that the additive utility functions approach imposes certain requirements on the measurement scales used for the items in a portfolio that should not be ignored, given the different results that can arise from the portfolio size effect. That is, the baseline correction may not be enough to avoid this problem.

Therefore, this paper suggests that additionally to the baseline adjustment, a ratio scale correction may be necessary, depending on the combination of values and constraints considered by the problem, since the portfolio size effect may occur in additive multicriteria portfolio analysis.

In addition, this work is an extension from previous studies mentioned in the text, such as: [1, 6, 7, 13], which are related to MCDM portfolio analysis in different perspectives and give support for the understanding of the problem presented here.

Also, it is essential to understand that this paper is not about the baseline specifically and the baseline does not exist in every problem, as explained by [12] and it is not present in this application either, as pointed out before.

References

1. Almeida, J.A., de Almeida, A.T., Costa, A.P.C.S.: Portfolio Selection of Information Systems Projects Using PROMETHEE V with C-Optimal Concept. Pesquisa Operacional (Impresso) **34**, 1–25 (2014)
2. Belton, V., Stewart, T.J.: Multiple Criteria Decision Analysis. Kluwer Academic Publishers, Boston (2002)
3. CDSID (Center for Decision Systems and Information Development). Additive Multicriteria Portfolio Analysis and linear value function with sensitivity analysis, web-based V1 (PU-A2MME-WT1) (2014). Software available by request in www.cdsid.org.br
4. Clemen, R.T., Smith, J.E.: On the choice of baselines in multiattribute portfolio analysis: a cautionary note. Decis. Anal. **6**(4), 256–262 (2009)
5. Cook, W.D., Green, R.H.: Project prioritization: a resource-constrained data envelopment analysis approach. Socio-Econ. Plann. Sci. **34**, 85–99 (2000)
6. de Almeida, A.T., Vetschera, R.: A note on scale transformations in the PROMETHEE V method. Eur. J. Oper. Res. **219**, 198–200 (2012)
7. de Almeida, A.T., Vetschera, R., Almeida, J.A.: Scaling issues in additive multicriteria portfolio analysis. In: Dargam, F., Hernández, J.E., Zaraté, P., Liu, S., Ribeiro, R., Delibašić, B., Papathanasiou, J. (eds.) EWG-DSS 2013. LNBIP, vol. 184, pp. 131–140. Springer, Heidelberg (2014)

8. Kleinmuntz, D.N.: Resource allocation decisions. In: Edwards, W., Miles, R.F., von Winterfeldt, D. (eds.) Advances in Decision Analysis, pp. 400–418. Cambridge University Press, New York (2007)
9. Liesiö, J., Punkka, A.: Baseline value specification and sensitivity analysis in multiattribute project portfolio selection. Eur. J. Oper. Res. **237**, 946–956 (2014)
10. Mavrotas, G., Diakoulaki, D., Caloghirou, Y.: Project prioritization under policy restrictions: a combination of MCDA with 0–1 programming. Eur. J. Oper. Res. **171**(1), 296–308 (2006)
11. Morton, A.: On the choice of baselines in portfolio decision analysis. Working Paper LSEOR 10.128, LSE Management Science Group (2010)
12. Morton, A.: Measurement issues in the evaluation of projects in a project portfolio. Eur. J. Oper. Res. **245**, 789–796 (2015)
13. Vetschera, R., de Almeida, A.T.: A PROMETHEE-based approach to portfolio selection problems. Comput. Oper. Res. **39**, 1010–1020 (2012)

DSS Technologies Underpinned by Business Intelligence and Knowledge Management

A Knowledge Based System for Supporting Sustainable Industrial Management in a Clothes Manufacturing Company Based on a Data Fusion Model

Gasper G. Vieira[1], Leonilde R. Varela[1(✉)], and Rita A. Ribeiro[2]

[1] Department of Production and Systems, School of Engineering, University of Minho, Campus de Azurém, 4800-058 Guimarães, Portugal
gaspar_vieira@hotmail.com, leonilde@dps.uminho.pt
[2] UNINOVA - CA3, 2829-516 Caparica, Portugal
rar@uninova.pt

Abstract. In this paper we propose a knowledge based system (KBS), based on smart objects and a data fusion model to support industrial management decision making applied to a clothes manufacturing enterprise. The management processes cover factory-production levels to higher decision-making levels. Therefore, the proposed KBS contributes to solving different kind of decision problems, including factory supervision, production planning and control, productivity management, real-time monitoring, and data acquisition and processing. The web access via different middleware devices and tools at different process levels, along with the use of integrated algorithms, decision methods, and smart objects, promote an optimized use of knowledge and resources. In this paper the proposed KBS is introduced and an example of its use is illustrated with an example of a clothes manufacturing resources selection, using the embedded dynamic multi-criteria fusion model.

Keywords: Knowledge based system · Industrial management decision making · Dynamic multi-criteria decision model · Manufacturing resources selection

1 Introduction

Nowadays, manufacturing enterprises are facing many challenges to respond to higher levels of production quality requirements, such as products quality requisites, manufacturing processes optimization and the manufacturing management processes itself, because globalization is forcing enterprises to promptly and accurately respond to requests arising from all over the world. An example is the necessity to accurately plan manufacturing resources usage, as the industrial manufacturing environment is no longer working isolated for satisfying its own manufacturing orders, but they also have to satisfy production needs from other outside orders. Therefore, industrial companies have to form strategic relationships with business partners to increase their responsiveness to market changes and to share resources more effectively and efficiently, through reliable decision support systems for supporting manufacturing management [1–4].

© Springer International Publishing Switzerland 2016
S. Liu et al. (Eds.): ICDSST 2016, LNBIP 250, pp. 113–126, 2016.
DOI: 10.1007/978-3-319-32877-5_9

Manufacturing management, at its higher level, involves defining strategies to connect people, processes, data (information), knowledge and decision-making. Further, in the current context of globalised markets, big quantities of more or less complex data has to be acquired and processed, to accurately make decisions in a daily basis, considering not just in-door information but also out-door one. Therefore, it becomes of utmost importance and necessity to integrate technology and knowledge for enabling accurate and timely responses to market requests [5–12].

In this paper we propose a knowledge-based system (KBS) for supporting manufacturing management decision-making, which includes a dynamic multi-criteria decision-making model (DMCDM) and a Data Fusion algorithm (FIF) [13–16] to ensure taking in consideration today's spatial-temporal global environments.

Moreover, the proposed KBS is linked with a set of smart objects, for collecting data at the machines and factory level, along with appropriate middleware technology and tools for supporting appropriate data storage and processing, thus improving decision-making processes. The concept of "Smart Objects" is well known and comes back from the late 1990's [17]. The main focus of the concept is on modeling interactions of smart virtual objects with virtual humans, agents, in virtual worlds".

This paper is organized as follows. In Sect. 2 an overview of dynamic decision-making and data fusion models is presented. Section 3 briefly describes the proposed KBS, including a general view of the integrated data acquisition and processing module, based on smart objects. Section 4, presents an industrial application example of the usage of the KBS for selecting manufacturing resources in a clothes manufacturing company. In Sect. 5 a brief literature review about related work is presented, and finally, Sect. 6 presents some conclusions and planned future work.

2   Background Context

In the literature, many interesting works can be found related to manufacturing management frameworks and approaches [1–12]. The research project in [1] addresses the emergence of interactive manufacturing management at three levels: sector, system and enabling technologies. More details about characteristics and capabilities of manufacturing frameworks will be further discussed on the description of the proposed KBS, Sect. 3.

As mentioned above, the KB System includes an embedded Dynamic Multi-criteria Decision Making model and Data Fusion algorithm. Classical MCDM is a technique widely used for selection problems [18–21] that assumes a fixed time frame where knowledge from past or future information is not employed to support more informed decisions.

The first step in the classical MCDM [18] is to identify the available alternatives, selecting relevant criteria to evaluate each alternative and develop the decision matrix based on the level satisfaction of each alternative for each criterion. This phase is usually called knowledge elicitation. The next phase is to aggregate the satisfaction values of criteria for each alternative to achieve a final per alternative (rating) so they can be ranked. Further, in the classic MCDM there is only one matrix reflecting the current status of the system, while in the Dynamic MCDM (DMCDM) model [15] at least two matrices must be considered, the historic matrix, which represents the situation in the past, and the current matrix, which represents the current status. At each period (time or iteration) the 2 matrices are combined and the result stored (updated historic data) for the next iteration. Details about the mathematical formulations for this dynamic decision making model can be seen in [13–15]. Here, we follow the idea from [14] of extending this dynamic model [15] with a "future knowledge matrix" representing the estimated future values for certain criteria to evaluate the alternatives of the current situation. The past status includes historical data and the future or predicted knowledge can be calculated either by using a forecasting model or using experts' knowledge. The future or predicted information could also be generated by negotiation and estimation [14]. In summary our approach will consider three different matrices: past, present and future, as suggested in [14].

To perform the calculations and obtain a final score of aggregating the three matrices we follow the data fusion process (FIF algorithm) described in [13, 16]. This data fusion process includes 5 steps: (1) normalize the criteria using fuzzification [8]; (2) filter uncertainty; (3) define criteria weights; (4) fusing information by aggregating criteria; (5) final ranking. The mathematical details of the combined models behind the five steps can be seen in [13–15].

3   Proposed KBS and Integrated Technologies

In this work we propose the integration of different manufacturing management functions, varying from the administration scope down to the operational level, passing through the production planning and control, as illustrated in Fig. 1.

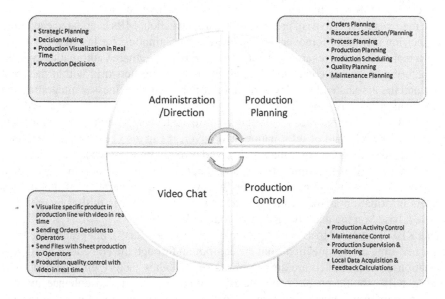

Fig. 1. Manufacturing management functions considered in the proposed KBS.

To manage the complexity of the management functions in real decision making problems there are different types of strategies to simplify the problem [19–21]. The common strategy is to consider the situation time-independent and model the problem in a static situation. In this case, many important factors will be disregarded and in some cases it will result in erroneous decisions. Furthermore, most tactical and strategic decision in companies require some thought and time, sometimes even undergoing internal negotiations between departments, to reach a final decision, i.e. these types of decisions are spatial-temporal dependent.

Although at the operational level there is the capability of gathering real data from the shop floor, through smart objects, in terms of strategic and even tactic planning, it is of upmost importance to be able to consider manufacturing scenarios and strategies, based on future data/information predictions, namely regarding information that arises from the outside of the factory, namely ad-hoc requests for using manufacturing resources or machines. These increasingly complex manufacturing management scenarios require using dynamic multi-criteria decision-making models, as many different kinds of information have to be considered. Therefore, in this work we consider a Dynamic Multi Criteria Decision Model (MCDM) [14, 15] along with a data fusion method [13] within the proposed KBS.

The proposed KBS and integrated technologies are mainly based on supervision equipment, data acquisition and processing devices and tools, including smart objects, as illustrated in Fig. 2.

Regarding past and current data it can be either collected automatically, through the smart objects interacting with manufacturing resources or manually through MR managers. This data is further inserted in the KBS and processed and analysed through the underlying DSS, which integrated the MCDM implementation.

The matrices related to past, present and future data about MRs is updated each time some decision is made. Moreover, it is updated in a continuous, for instance daily and also in a real-time basis, regarding the information that is gathered automatically from manufacturing resources connected through the underlying network, within the local factory and the associated factories or outdoor collaborating businesses integrating the extended enterprise environment.

Future data can be predicted at different confidence levels through prevision models appropriate for each kind of information. Moreover, it can also be inserted manually, for instance, regarding new information received directly by local resource managers.

The three matrices regarding past, present and future information are associated through the MCDM in a dynamic and iterative way, as each iteration of this model, considers information arising from these three matrices, which is further merged and processed, based on the underlying data fusion method [13].

This KBS is intended to act as a "System-as-a-Service" (SaaS), integrating the services for real-time data acquisition from the equipment through the embedded intelligent information devices, which are smart objects in a clothes factory environment. The proposed KBS, should enable better decision-making support and enhance human-human and human-machine interactions, by means of the integration of information and its processing functions.

Our illustrative example for the proposed integrated KBS, implements an application in the context of a clothes factory in Portugal. Figure 2 illustrates the general view about the principal entities and corresponding main interactions considered for data acquisition and processing within the clothes manufacturing environment, where the manufacturing resources in the production line of the clothes factory play a fundamental role. The Manufacturing Resources (MR) can be any provider of any service, machine tool type, human agents as service providers (designers, managers, machine operators, planners, schedulers, drivers, vendors, and others), computing resources, software, among others. The MR receives the orders from internal or external 'clients' and then negotiations are triggered (for example via chat, video conferencing, or email). After the approval of the order, the Resource establishes direct relationship with the client and executes the production order. As stated above, the Resource may give permission for the client to see the production order to be executed and may allow the client to control the use of distance (when the resource is a machine, a computer, or software), either from the control room, the PC, or from a mobile device, and even remotely operated [12].

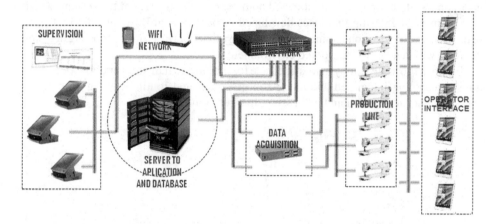

Fig. 2. Integrated technologies in a clothes factory environment.

In general, the integrated technologies include equipment for interactive monitoring systems and product design services that integrate four environments: (1) computer aided design, (2) product data repository with embedded system for decision making (for accessing all relevant data, such as current, historic or forecasted data as well as data analysis) from the equipment in use, and equipment operation services that integrate the following environments [9]:

(1) Real-time equipment data that provides all relevant data, actual and historic as well as forecasted data and corresponding data analysis and suggestions, necessary for the production management;

(2) Management environment information for monitoring, planning, scheduling and controlling production activities;

(3) An interactive environment for supporting management – services;

(4) A 'Cloud' infrastructure, necessary to provide: (a) infrastructure for the manufacturing system applications – of all three types of resources: material, manufacturing resources, information processing and other resources (i.e. computational resources), and knowledge resources – in the form of IaaS - Infrastructure as a Service; (b) platform for the manufacturing system applications in the form of PaaS - Platform as a Service; (c) manufacturing system software 'business' applications in the form of SaaS - Software as a Service.

The KBS has been implemented in Visual Basic (VB) language to prototype the clothes application for this paper. The VB language enables easy development; as well as user friendly interfaces for data visualization and processing. Data and corresponding processing approaches and applications are managed in different places, which is of particularly importance in the context of a decentralized manufacturing scenario. For instance, if these globally distributed manufacturing environments have their own schedulers developed by different IT stakeholders; their schedules cannot be visualized by a common viewer without particular adapting programs. This actually causes a huge effort on a system's implementation, and both the cost and the risk of the system, which will be increased. Using interfaces developed through Visual Basic enables end users to have a personalized scheduling viewer, among other interfaces for supporting decision making, within a whole networked environment.

Figure 3 illustrates the system's interface for the clothes industrial application, which aims at enabling integrated and automatic processes and routines, along with corresponding data acquisition and processing, in a real-time basis.

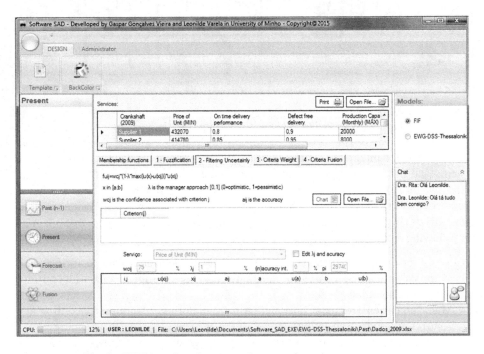

Fig. 3. KBS interface for supporting manufacturing management.

Specifically, Fig. 3 depicts the interface for supporting manufacturing resources evaluation and selection, using the described dynamic multi criteria decision model (DMCDM) [14]. This interface provides flexibility for visualizing distributed plans everywhere through the Internet. Therefore, the proposed data representation and processing model can be seen as a general modeling schema, for problem data specification and processing for enabling to better supporting decisions at different decision levels intra and/or inter factories and stakeholders.

Moreover, one important aspect of the proposed KBS is its capability for enabling to acquire real data from the machines and other manufacturing resources, at the factory level, through smart objects technology, in a precisely and real-time basis [17, 22–25]. The smart objects have the capability to collect and store data in real time, to identify themselves and to make decisions, in a automatic and autonomous way, thus they play a crucial role in terms of real-time management functions, for supporting manufacturing, as they enable to update data to the second, instead of the traditional reports, which take sometimes days or even more time periods to enable to use updated data for manufacturing decision-making support.

The proposed KBS enables either to insert information about past, present and future data predictions regarding manual entries or directly in the database through automatic data insertion, namely data that is driven from the execution of the dynamic decision making model, once decisions are made. For this purpose the KBS includes some production rules that enable to filter relevant information, regarding positive decisions that are made, each time some manufacturing resource or business is selected for accomplishing a given manufacturing order. An illustrative example is provided in Fig. 4 representing a production rule's pseudo-code from the KBS, which enables to store the data associated to MRs that have Production Costs (PC) that are half the value of a given product's price and which have a Quality Score (QS) higher than 70 % or a Number Of Complains (NOC) which is less than 4.

> If PC < 0,5Price
>   and QS > 70% or NCO < 4
> then Store MRi's data

Fig. 4. Pseudo-code example of a production rule for automatic knowledge acquisition.

Through this kind of production rules the KBS is able to "learn" or update the past, present and future data matrices based on current selection of the manufacturing resources. This means that each time a MR (or business) is selected, and this decision is evaluated by the KBS, the KBS will automatically update its knowledge based on several distinct production rules, such as the one illustrated above, according to each specific context and requirements. This step is actually a very vital characteristic of the KBS, which enables it to capture data and support decision-making on a dynamic basis. Another important form through which the KBS can learn from the data is related to the manual data entries, for instance, regarding information that is gathered and introduced by system operators, each time new information arises, namely regarding data updates or previsions from stakeholders, for instance, each time future production and/or service prices' tables are announced or predicted by the manufacturing resources or services providers managers.

Furthermore, the smart objects are programmed with the proposed functions in order to enable to manage, in real time, machines and products, sending accurate, timely and reliable information, to workers responsible for production planning and control, machining, assembling and maintenance. The functions are included in the modules, where each one represents a type of data that a smart object can capture. Moreover, there is a higher level requirement about a need to distribute the whole data by the existing hierarchical levels in the enterprise, since each job title requires singular responsibilities and decision making, and the smart objects' architecture allows crossover through different technological levels.

4   Industrial Application Example of the MCDM Module

Here we present and illustrative example of selecting a manufacturing resource for producing a given product, which is an adaptation from the example in [14].

Let us consider six alternative manufacturing resources (MR1, MR2, ..., MR6) available from different factories, which can be used for producing a given set of products. Six criteria were chosen for both past and future data evaluation, and a different set of five criteria was chosen for present data evaluation, as shown in Tables 1 and 2, respectively.

The present criteria are the following: Production Cost (PC); Estimated Delivery Time (EDT); Lead Time (LT), which refers to the amount of days needed before the product processing starts (quoted); Remote Operation Capability (ROC), and Distinct Product Batches (DPB). The DPB criterion provides information about how flexible a given resource is, based on its processing potential, in terms of enabling different kind of products to be processed simultaneously in the same resource, at the same time or within the same production order. The historical and future information is evaluated by the six criteria: Production Cost, per hour (PC); On-time Delivery (OD); Daily Delay Penalization (DDP), based on the number of days orders were delayed; Quality Score (QS), about work delivered; Number of Complains per Order (NCO); and Portfolio Score (PS).

Table 1. Present data about candidate manufacturing resources

Manufacturing Resources	Production Cost, PC	Estimated Delivery Time, EDT	Lead Time, LT	Remote Operation Capability, ROC	Distinct Product Batches, DPB
Resource 1 - MR1	325	4	2	1	8
Resource 2 - MR2	560	3	3	0,95	7
Resource 3 - MR3	450	5	2	0,9	4
Resource 4 - MR4	375	2	4	1	6
Resource 5 - MR5	290	6	3	0,8	3
Resource 6 - MR6	340	6	2	0,9	7

Table 2. Past and future data about candidate manufacturing resources (there is no past data for MR6 and therefore no future data also.)

Manufacturing Resources	Production Cost, PC		On-time Delivery, OD		Daily Delay Penalization, DDP		Quality Score, QS		Number of Complains per Order, NCO		Portfolio Score, PS	
	past	future	past	future	past	future	past	future	past	future	past	future
MR1	60	0,95	10	0,85	2	0,9	60	0,95	10	0,85	2	0,9
MR2	55	0,85	8	0,95	1	0,9	55	0,85	8	0,95	1	0,9
MR3	45	0,9	4	0,96	1	0,85	45	0,9	4	0,96	1	0,85
MR4	60	0,75	12	0,9	3	0,95	60	0,75	12	0,9	3	0,95
MR5	65	0,9	16	0,98	1	0,95	65	0,9	16	0,98	1	0,95
MR6												

Using the values associated with each criterion for the three types of matrices (past, present and future) we performed the data fusion process with FIF algorithm [13], described in Sect. 2 (see also the steps in the interface, Fig. 3): (1) normalize the criteria using fuzzification; (2) filter uncertainty; (3) define criteria weights; (4) fusing information by aggregating criteria; (5) final ranking. Illustrating, Table 4 displays the results obtained for the historical matrix by performing the initial data preparation process.

Table 3. Normalized historical data with corresponding relative importance (Legend of Table 3: (a) fuij are the normalized and filtered values from Table 2 for each criteria (step 1 and 2 of FIF algorithm); (b) L(fuij)) – are the criteria relative importance (step 3 of FIF), which depend on the satisfaction level of the criteria and on the assigned relative importance (see 3rd line on Table 3 very important (VI); important (I); Average importance (Av)). For details on how to calculate Table 3 values see [13, 14].)

Criterion	PC		OD		DDP		QS		NCO		PS	
Weights	fuij	L(fuij)	fuij	L(fuij)	fuij	L(fuij)	fuij	L(fuij)	fuij	L(fuij)	fuij	L(fuij)
	VI		I		I		Av		I		Av	
MR1	0,25	0,699	0,9	0,96	0,455	0,781	0	0,479	0	0,479	0,248	0,489
MR2	0,5	0,799	0,412	0,764	0,612	0,844	0,558	0,658	0,2	0,543	0,248	0,489
MR3	1	1	0,614	0,845	0,95	0,98	0,613	0,676	0,2	0,543	0	0,419
MR4	0,25	0,699	0	0,599	0,301	0,719	0,282	0,569	0	0,479	0,9	0,672
MR5	0	0,599	0,614	0,845	0	0,599	0,85	0,752	0,2	0,543	0,9	0,672
MR6												

After normalizing, filtering and determining the relative importance for each criterion we can calculate the rating for each criteria, per alternative. Illustrating, to determine the rating for criteria PC of alternative MR1 with past/historical information from Table 3 we have:

(a) first we get the rating for each criterion, such as exemplified for PC:

$$R(PC): 0.25 * 0.699/3.887 = 0.045$$

where,

$$Sum(L(fuij)) = 0.699 + 0.960 + 0.781 + 0.479 + 0.479 + 0.489 = 3.887$$

(b) second we calculate the rating for alternative MR1,

$$R(MR1) = 0.045 + 0.2223 + 0.0914 + 0 + 0 + 0.0312 = 0.389$$

After obtaining the ratings for each matrix, past, present and future, the dynamic spatial-temporal process can be used [13, 14] for obtaining the final rating for all candidate businesses at time t by aggregating the three matrices. Table 4 illustrates the final rating of the dynamic process using, again, the weighted average aggregation with weighting functions [13] where the relative weights depend on the values satisfaction. The last column of Table 4 contains the final vector ratings for all candidate alternatives.

Table 4. Final ratings

Manufacturing Resource	Historical		Present		Future		Final rating
	rating	*Weight (I)*	*rating*	*Weight (VI)*	*rating*	*Weight (Av)*	
MR1	0,39	0,68	0,891	0,946	0,318	0,581	**0,586**
MR2	0,446	0,7	0,604	0,802	0,529	0,649	**0,53**
MR3	0,666	0,779	0,377	0,689	0,421	0,614	**0,498**
MR4	0,309	0,651	0,76	0,88	0,361	0,595	**0,51**
MR5	0,467	0,707	0,391	0,696	0,301	0,576	**0,392**
MR6	0	0,539	0,564	0,782	0	0,479	**0,245**

Illustrating again for MR1, we determined the final score for each alternative using the dynamic model [13, 14] as follows,

$$Sum(L(uij)) = 0.680 + 0.946 + 0.581 = 2.207$$
$$Final\ score = (0.680/2.207) * 0.390 + (0.946/2.207) * 0.891 + (0.581/2.207) * 0.318$$
$$= 0.12 + 0.382 + 0.0837 = 0.586$$

After analysing the final scores obtained we can conclude that the products processing task should be assigned to the top ranked candidate manufacturing resource, "MR1", which displayed a final score of 0.586.

5   Related Work and Contribution

In recent years, the rapid development in information technology and in particular regarding internet technology turns the complex interaction and interoperability problems quite simpler to solve. However, despite the many advantages arising from these recent technologies, there are still some issues that have to be overcome particularly regarding collaboration levels integration and corresponding information processing and decision-making support that has to be assured within distributed manufacturing networks, which require dynamic and real-time decision making capabilities. Therefore, it is necessary to propose and develop frameworks, architectures, and methods by combining them with the current collaborative network models to compete within the global market scenario.

In this direction, during the last decade, the rapid development of NICT (New Information and Communication Technologies), with special emphasis on the advancement of RFID technology, Bluetooth and Wi-Fi, allowed the development of new production systems tools with traceability, visibility and interoperability in real time facilitating planning and shop floor control [26–28]. This means that any manual activity, time consuming and prone to errors associated with data collection and processing is able to be reduced or even eliminated, since the capture and processing of information that takes place in real time, thus allowing a more rapid and accurate decision-making [26].

The concept of smart object plays a key role in this new generation production systems to explore the integration of physical objects with the technologies outlined above, in order to "acquire" a set of said "smart" properties.

That is, a product throughout the supply chain, is not only a physical good, but a key element in the information infrastructure, through interaction with other products, processes and stakeholders in this same chain. To offer companies a new paradigm of interaction with their products and manufacturing resources are expected significant gains in operational efficiency. Automatic monitoring and context perception enable one best performance of information systems, such as Supply Chain Management, the Enterprise Resource Planning and Warehouse Management Systems, as they are no longer fed by outdated information [21].

Frequently, the major problem associated with the interaction of the smart objects is that sometimes they use different communication protocols, raising problems of compatibility and synchronization of information [27].

There is a variety of interesting and more or less closely related decision-making tools available, for instance [26–28]. Although, our proposed decision-making system provides some extensions, regarding the integration of several distinct technologies, among which are also included, smart objects for local and remote data collection and processing directly from manufacturing resources or managers. Moreover, our proposed system is based on a dynamic MCDM, which enables to integrate and process past, present and future information, regarding a variable set of criteria, according to each particular request arising in the context of different manufacturing management decision levels and requests as described before through an example of use occurring in the context of a clothes manufacturing environment where the system is being implemented and tested.

6 Conclusion

This paper described a knowledge based system (KBS) for supporting industrial management decision making in a clothes factory. The proposed KBS integrates a set of technologies which enable to directly interact with machines and equipment in the shop floor for data acquisition and its subsequent processing for supporting industrial management decision making (IMDM). One important tool for this IMDM is a module that integrates a combined dynamic multi criteria decision model (DMCDM) and data fusion model (FIF algorithm) for supporting evaluation and selection of manufacturing resources (MR), from a pool of MRs put available either locally in the clothes factory shop floor or in another associated factory. An illustrative example of the application of the MCDM plus FIF was presented in the paper for clarifying its utility for the proposed KBS.

Moreover, the effectiveness of the integrated technologies and approaches was briefly described and illustrated through the application example provided, namely regarding smart objects, which play a very important role in the proposed KBS, namely at the shop floor level for the data acquisition and local data processing. An important aspect of the KBS is that data can be generated and visualized by computers and other devices, including the smart objects, in appropriate and distinct ways and it is also important to notice that the data representation schema is general for distinct kind of manufacturing requisites.

For being able to fully implement the overall characteristics of the proposed KBS, there is still some need of further work to implement additional functionalities, as for instance for fully implementing the data fusion algorithm (FIF) and then tests and validation on the complete KBS also have to further continuing to take place.

Acknowledgments. This work was supported by FCT "Fundação para a Ciência e a Tecnologia" under the program: PEST2015-2020, reference: UID/CEC/00319/2013.

References

1. Shi, Y., Fleet, D., Gregory, M.: Global Manufacturing virtual network (GMVN): a revisiting of the concept of after three years fieldwork. J. Syst. Sci. Syst. Eng. **4**(12), 432–448 (2003)
2. Varela, M.L.R., Putnik, G.D., Cruz-Cunha, M.M.: Web-based technologies integration for distributed manufacturing scheduling in a virtual enterprise. Int. J. Web Portals **4**(2), 19–34 (2012). doi:10.4018/jwp.2012040102)
3. Arrais-Castro, A., Varela, M.L.R., Putnik, G.D., Ribeiro, R.A.: Collaborative network platform for multi-site production. In: Hernández, J.E., Zarate, P., Dargam, F., Delibašić, B., Liu, S., Ribeiro, R. (eds.) EWG-DSS 2011. LNBIP, vol. 121, pp. 1–13. Springer, Heidelberg (2012). doi:10.1007/978-3-642-32191-7_1
4. Carvalho, J.B., Varela, M.L.R., Putnik, G.D., Hernández, J.E., Ribeiro, R.A.: A web-based decision support system for supply chain operations management - towards an integrated framework. In: Dargam, F., Hernández, J.E., Zaraté, P., Liu, S., Ribeiro, R., Delibašić, B., Papathanasiou, J. (eds.) EWG-DSS 2013. LNBIP, vol. 184, pp. 104–117. Springer, Heidelberg (2014). doi:10.1007/978-3-319-11364-7_10
5. Lee, J.Y., Kim, K.: A distributed product development architecture for engineering collaborations across ubiquitous virtual enterprises. Int. J. Adv. Manuf. Technol. **33**(1–2), 59–70 (2006). (Springer-Verlag London Limited)
6. Appleton, O., et al.: The next-generation ARC middleware. Ann. Telecommun. **65**, 771–776 (2010)
7. Varela, L.R., Ribeiro, R.A.: Evaluation of Simulated Annealing to solve fuzzy optimization problems. J. Intell. Fuzzy Syst. **14**, 59–71 (2003)

8. Putnik, G.: Advanced manufacturing systems and enterprises: cloud and ubiquitous manufacturing and an architecture. J. Appl. Eng. Sci. **10**(3), 229, 127–134 (2012)
9. Putnik, G.: Advanced manufacturing systems and enterprises: cloud and ubiquitous manufacturing architecture. J. Appl. Eng. Sci. **10**(3), 127–229 (2012). doi:10.5937/jaes10-2511
10. Cunha, M.M., Putnik, G.D.: Market of resources as a knowledge management enabler in VE. In: Jennex, M.E. (ed.) Knowledge Management: Concepts, Methodologies, Tools, and Applications, pp. 2699–2711 (2008)
11. Putnik, G.D., Castro, H., Ferreira, L., Barbosa, R., Vieira, G., Alves, C., Shah, V., Putnik, Z., Cunha, M., Varela, L.: Advanced Manufacturing Systems and Enterprises – Towards Ubiquitous and Cloud Manufacturing, University of Minho, School of Engineering, LabVE (2012)
12. Ribeiro, R.A., Falcão, A., Mora, A., Fonseca, J.M.: FIF: A Fuzzy information fusion algorithm based on multi-criteria decision making. Knowl.-Based Syst. J. **58**, 23–32 (2013). doi:10.1016/j.knosys.2013.08.032
13. Jassbi, J.J., Ribeiro, R.A., Varela, L.R.: Dynamic MCDM with future knowledge for supplier selection. J. Decis. Syst. 232–248 (2014). doi:10.1080/12460125.2014.886850, Taylor & Francis
14. Mora, A.D., Falcão, A.J., Miranda, L., Ribeiro, R.A., Fonseca, J.M.: A fuzzy multicriteria approach for data fusion (chap. 7). In: Fourati, H. (ed.) Multisensor Data Fusion: From Algorithms and Architectural Design to Applications, pp. 109–123. CRC Press, Boca Raton (2016). ISBN 978-1-4822-6374-9
15. Campanella, G., Ribeiro, R.A.: A Framework for dynamic multiple criteria decision making. Decis. Support Syst. **52**(1), 52–60 (2011). doi:10.1016/j.dss.2011.05.003
16. Angels, R.: RFID technologies: supply-chain applications and implementation issues. Inf. Syst. Manage. **22**(1), 51–65 (2005)
17. Ng, S.T., Skitmore, R.M.: CP-DSS: decision support system for contractor prequalification. Civil Eng. Syst. Decis. Mak. Probl. Solving **12**(2), 133–160 (1995)
18. Triantaphyllou, E.: Multiple Criteria Decision Making Methods: A Comparative Study. Kluwer Academic Publishers, Boston (2000)
19. Figueira, J., Greco, S., Ehrgott, M.: Multiple Criteria Decision Analysis: State of the Art Surveys. International Series in Operations Research & Management Science, vol. 78. Springer, New York (2005)
20. De Boer, L., Labro, E., Morlacchi, P.: A review of methods supporting supplier selection. J. Purchasing Supply Manage. **7**(2), 75–89 (2001). ISSN 1478-4092
21. Bajic, E.: A service-based methodology for RFID-smart objects interactions in supply chain. Int. J. Multimedia Ubiquitous Eng. **4**(3), 37–54 (2009)
22. Beigl, M., Gellersen, H.: Smart-its: an embedded platform for smart objects. In: Proceedings of the Smart Objects Conference 2003 (sOc) (2003)
23. Huang, G.Q., Zhang, Y., Chen, X., Newman, S.T.: RFID-based wireless manufacturing for walking-worker assembly islands with fixed-position layouts. Robot. Comput. Integr. Manuf. **23**(4), 469–477 (2008)
24. Zhekun, L. Gadh, R., Prabhu, B.: Applications of RFID technology and smart parts in manufacturing. In: Proceedings of the ASME 2004 International Design Engineering Technical Conference and Computers and Information in Engineering Conference (2004)
25. Marcelo, K., Thalmann, D.: Modeling objects for interaction tasks. In: Arnaldi, B., Hégron, G. (eds.) Computer Animation and Simulation 1998. Eurographics, pp. 73–96. Springer, Vienna (1998)

26. Huang, G.Q., Zhang, Y., Jiang, P.: RFID-based wireless manufacturing for real time management of job shop WIP inventories. Int. J. Adv. Manuf. Technol. **36**(7–8), 752–764 (2008)
27. Zhang, Y., Huang, G.Q., Qu, T., Ho, O., Sun, S.: Agent-based smart objects management system for real-time ubiquitous manufacturing. Robot. Comput. Integr. Manuf. **27**(3), 538–549 (2011)
28. Zhang, Y., Huang, G.Q., Qu, T., Sun, S.: Real-time work-in-progress management for ubiquitous manufacturing environment. In: Li, W., Mehnen, J. (eds.) Cloud Manufacturing. Springer Series in Advanced Manufacturing, pp. 193–216. Springer, London (2013)

Knowledge Management as an Emerging Field of Business Intelligence Research: Foundational Concepts and Recent Developments

Sean B. Eom[✉]

Accounting Department, Southeast Missouri State University, Cape Girardeau, USA
sbeom@semo.edu

Abstract. A number of prior studies have been conducted to assess the extent of progress within these stages in the BI area. Among them, a study of Eom [1] has provided bibliometric evidence that the decision support system has made meaningful progress over the past three and a half decades (1969–2004). The primary data for this study were gathered from a total of 498 *citing* articles in the BI/DSS area over the past eight years (2005–2012). This study, based on author cocitation analysis (ACA), presents two important findings. First, the empirical consensus of BI researchers reveals that the focus of business intelligence research is shifting to knowledge management and data mining. Second, since ACA is a supporting quantitative tool that must be used with further qualitative analysis of bibliographic data, we examined the foundational concepts of knowledge management provided by the most influential scholars and their most frequently cited publications.

Keywords: Business intelligence · Decision support systems · Data mining · Knowledge management · Informetrics · Author cocitation analysis · Multidimensional scaling

1 Introduction

Over the past five decades, the area of decision support systems has made a significant progress toward becoming a solid academic discipline. For DSS to become a coherent and substantive field, a continuing line of research must be built on the foundation of previous work. Without it, there may be good individual fragments rather than a cumulative tradition (Keen, 1980).

A number of prior studies have been conducted to assess the extent of progress within these stages in the DSS area. Among them, a study of Eom [1] has provided bibliometric evidence that the decision support system has made meaningful progress over the past four and a half decades (1969–2004). His study applies factor analysis, cluster analysis, and multidimensional scaling of an author cocitation frequency matrix derived from a bibliometric database file. The results of his bibliometric analysis shed some lights on the development and evolution of DSS research over the period of 1969 through 2004.

© Springer International Publishing Switzerland 2016
S. Liu et al. (Eds.): ICDSST 2016, LNBIP 250, pp. 127–136, 2016.
DOI: 10.1007/978-3-319-32877-5_10

In light of these developments, this study focuses on examining the intellectual structure of business intelligence (BI) research during the period of 2005 and 2012. BI is an umbrella term that combines transaction databases, data warehouses, and data analytical tools such as structured query languages, online analytical processing, data mining, visualization, etc. [2]. ABI/Inform database retrieved 25,612 peer reviewed publications since 2005 with the descriptor, "knowledge management". The authors and their publications representing the knowledge management cluster in this paper are definitely the cream of the crop. They are the most frequently cited authors and their publications.

The empirical consensus of BI researchers reveals that knowledge management is an emerging field of business intelligence research appeared recently. The longitudinal analysis of more than 3 decades of decision support systems research [1, 3, 4] revealed that knowledge management and data mining are two emerging business intelligence research subspecialties. No other research tools other than ACA can provide a bird's-eye view of the BI research with unobtrusive and objective characteristics. Second, since ACA is a supporting quantitative tool that must be used with further qualitative analysis of bibliographic data, we examined the foundational concepts of knowledge management provided by the most influential scholars and their most frequently cited publications. Specifically, the ACA based research aims to focus on only knowledge management answer the following questions,

1. What are foundation concepts in knowledge management?
2. What are recent development in knowledge management?

The organization of this paper is as follows. Section 2 describes the bibliometric data used for this study. Section 3 is concerned with research methodology. Section 4 presents results of the current study. Section 5 summarizes the findings and discuss the implication of the current study for future BI research. The final section is the conclusion.

2   Data

The primary data for this study were gathered from a total of 498 *citing* articles in the BI area over the past eight years (2005–2012). BI is an umbrella term that encompasses a wide range of information systems that discover, collect, store, use, and distribute information. The BI system consists of the data warehousing environment, the business analytics environment, and the performance monitoring system. The business analytics environment is responsible for producing queries and reports as well as discovering useful information, and supporting decision makers and top management to make effective decisions.

The citing articles include articles related to DSS, ESS/EIS, data warehouse, data mining, business performance monitoring systems, and other business analytical tools such as OLAP and SQL. We applied rigorously the selection criteria explained in the data section to screen all articles published in the following 11 major information systems, operations research and management science journals. They are *Decision Sciences, Decision Support Systems, European Journal of Operational Research,*

Information & Management, Information Systems Research, Interfaces, Journal of Management Information Systems, Management Science, MIS Quarterly, and Omega.

3   Research Methodology

This study uses author cocitation analysis (ACA). ACA is a technique of bibliometrics that applies quantitative methods to various media of communication such as books, journals, conference proceedings, and so on. Citation analysis is often used to determine the most influential scholars, publications, or universities in a particular discipline by counting the frequency of citations received by individual units of analysis (authors, publications, etc.) over a period of time from a particular set of citing documents. However, citation analysis cannot establish relationships among units of analysis. ACA is the principal bibliometric tool to establish relationships among authors in an academic field and thus can identify subspecialties of a field and how closely each subgroup is related to each of the other subgroups.

The correlation matrix derived from the co-citation matrix of 87 authors was used as an input to the PROC MDS procedure of SAS Enterprise guide 4.2. Multidimensional scaling (MDS) is a class of multivariate statistical techniques/procedures to produce two- or three-dimensional pictures of data (geometric configuration of points) using proximities among any kind of objects as input. The purposes of MDS are to help researchers identify the "hidden structures" in the data and visualize relationships among/within the hidden structures to give clearer explanations of these relationships to others [5, 6]. Three SAS procedures (MDS, PLOT, and G3D) are necessary to convert the author cocitation frequency matrix to two- or three-dimensional pictures of data.

The distance matrix should be converted to a coordinate matrix. The coordinate matrix is used to produce two-dimensional plots and annotated three-dimensional scatter diagrams. Readers are referred to [7, 8] for further details of the methodologies used in this study.

4   Results of Multidimensional Scaling (MDS)

Figure 1 presents a two dimensional MDS map, which is the projection of the first and second planes presented in the three-dimensional MDS map. It shows the big picture of inter-cluster relationships. The placement of authors on the center of the MDS map means that those authors are linked with a substantial portion of the author set, with relatively high correlations. Placement near the periphery represents a more focused linkage. This is illustrated by the data mining researchers (Vapnik, Quinlan, Setiano, Witten, Frank, Baesens, etc.) located in the lower right hand area and knowledge management researchers (Nonaka, Strauss, Agote, Davenport, Prusak, Holland, etc.) located in the upper-right hand area. The location of the MCDM scholars (Saaty, Keeney and Raiffa, Roy, and Wallenius), in the center of the lower MDS map seems to indicate that MCDM is a major contributing discipline in that it connects with the data mining area and multiple criteria DSS scholars such as Jark, Bui, and Kertsten.

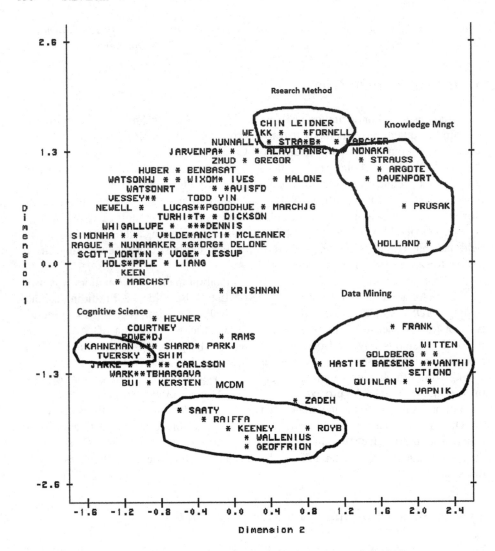

Fig. 1. Two dimensional MDS map of BI research (2005–2012)

Figure 1 represents several clusters of BI research sub-specialties and reference disciplines during the period of 2005 and 2012. The clusters in Fig. 1 include group support systems, foundations, research method, knowledge management, data mining, organization science, cognitive science, and multiple criteria decision making. Naming each cluster is based on the nature of most frequently cited documents of authors in any cluster. Readers are referred to Eom [9] for the overview of each of business intelligence subspecialties.

4.1 Foundational Concepts of Knowledge Management

The cluster at the top right-hand side of Fig. 1 seems to define knowledge management as it represented by Argote et al. [10–12], Nonaka [13, 14], Davenport et al. [15] and Prusak [16], Alavi [17] and Leidner [18]. Knowledge management can be defined as "the systematic and active management of ideas, information, and knowledge residing in an organization's employees" [2, p. 474]. The core of knowledge management concepts are the knowledge management process which is comprised of knowledge creation, storage/retrieval, dissemination, and application.

4.1.1 The Knowledge Management Process

Argote et al. [10] provides an integrative two dimensional framework for organizing the literature on knowledge management: the process (knowledge creation, retention, and transfer) and entities involved in knowledge management (individual, group, organization). Argote and Ingram [11] argues that the creation and transfer of knowledge embedded in the interactions of people, tools, and tasks provides a basis for competitive advantage in firms. Alavi and Leidner [18] reviewed conceptual foundations of knowledge management and knowledge management systems to present a systematic framework which is comprised of four sets of knowledge processes: creation, storage/retrieval, transfer, and application. Further, the potential role of IT and research issues that can arise on each activity in the framework are discussed [18].

4.1.2 The Knowledge Creating Process

Nonaka [13] proposes a paradigm for managing the dynamic aspects of organizational knowledge creating processes. Organizational knowledge is created through an ongoing four modes of knowledge creation: socialization, internalization, externalization, and combination. Nonaka argued that new knowledge is developed by individuals, but organizations play a critical role in articulating and amplifying that knowledge.

4.1.3 Critical Success Factors of Knowledge Management Projects

Davenport et al. [15] identified several characteristics of successful knowledge management projects such as link to economic performance, organizational IT infrastructure, knowledge-friendly culture, clear purpose, multiple channels for knowledge transfer, and senior management support.

4.2 Recent Developments in Knowledge Management

The DSS researchers focused on the roles of IT in knowledge creation, applying the four modes (internalization, externalization, socialization, and combination) of knowledge creation identified by Nonaka [13]. Argote and Ingram [11] illuminated that the creation and transfer of knowledge internally while preventing its external transfer to competitors are a basis for competitive advantage in firms, because people within an organization are more similar than people between organizations.

During the past decade, IS researchers developed a method for ontology-based empirical knowledge representation and reasoning. Others developed a model that explains how individuals' perceptions of three primary validation process characteristics (duration, transparency, and restrictiveness) impact their perceptions of repository knowledge quality and their contribution behaviors. The key papers in the knowledge management cluster are cited to further develop and extend those concepts into the following areas of knowledge management over the past 10 years (2005–2015). The ACA data covers only up to 2012. But we include the literature up to 2015.

4.2.1   Creating and Capturing Knowledge

This topic is also known as knowledge discovery in data bases (KDD). KDD is a process to identify, search for, and extract actionable information from data warehouses.

- Guidelines for designing visual ontologies to support knowledge identification is suggested [19].
- A novel way of capturing correlations across the entire data warehouse to acquire more knowledge from corporate data warehouses [20].
- Development of a method of ontology-based empirical knowledge representation and reasoning, which adopts OWL (Web Ontology Language) to represent empirical knowledge in a structural way [21].
- Development of Data mining tools based on a new approach to classification based on an extended association rule mining [22].
- A comparative analysis of machine learning techniques [23].
- Application of a hybrid of genetic algorithm and particle swarm optimization algorithm for order clustering [24].

4.2.2   Storing Knowledge

- Deriving knowledge representation guidelines by analyzing knowledge engineer behavior [25].
- Collective taxonomizing: A collaborative approach to organizing document repositories [26].

4.2.3   Disseminating Knowledge

- A field study of how organizations share and disseminate knowledge to improve their productivities as well as team performance [27].
- Transactive memory systems in organizations [28].
- Knowledge sharing using an agent-based simulation model [29].
- key issues in designing and using knowledge flow networks [30].
- Creating and using knowledge networks and development of a knowledge framework that represents knowledge fragments that need to be integrated [31].

- Understanding of continued knowledge sharing via the KMS [32].
- Using domain-specific knowledge in generalization error bounds for support vector machine learning [33].

4.2.4 Knowledge Management System Design

- Development of a comprehensive model to empirically examines how intellectual capital (IC) and knowledge management (KM) affect each other, and their consequences [34].
- KMS design principles and their applications to management of knowledge concerning pediatric bipolar disorder [35].

4.2.5 Data Mining as a Knowledge Creation Tool

Data mining is concerned with the process of discovering knowledge from large databases by means of various tools and techniques developed in statistics, OR/MS, and artificial intelligence. It seeks to identify three major types of patterns: prediction, association, and clustering. Predicting future occurrences of certain events can be achieved by building the classification model or regression model by analyzing the historical data in the database and build the models that can predict future behavior of entities such as customers or stock prices, or temperatures. Building classification/regression models involves the use of many tools/algorithms such as decision tree, artificial neural networks, support vector machines, genetic algorithms, rough sets, multiple regressions, etc. support vector machines are built on the statistical learning theory of Vapnik [36–38]. Clustering is an important data mining process of grouping entities or events into segments based on similarity of their attributes. The clustering process utilizes many techniques and algorithms such as self-organizing maps, k-means, etc.

Most authors appeared in the data mining cluster in Fig. 1 are the author of two classic books on support vector machines [36–38], the proposer of the C4.5 algorithm for extracting classification rules from decision trees [39–41]. In the same line of research, Baesens et al. applied neural network rule extraction and decision tables for credit-risk evaluation [42]. Hastie et al. suggested a statistically-oriented approach to clustering [43]. Witten and Frank's book discussed most of data mining tools/algorithms except neural networks and genetic algorithms. The tools/algorithms include decision trees, classification, and association rules, support vector machines, clustering, and regression [44].

Data mining is applied to many areas including credit data analysis. [45], predicting going concern opinion [46], developing pharmaceutical and manufacturing processes, predicting the length of hospital stay of burn patients, comparatively analyzing machine learning techniques for student retention management.

5   Conclusion

The unique contributions of this study is as follows. First, the empirical consensus of BI researchers reveals that knowledge management is an emerging field of business intelligence research appeared recently. This is a significant finding of this research. No other research tools can provide a bird's-eye view of the BI research with unobtrusive and objective characteristics. Second, since ACA is a supporting quantitative tool that must be used with further qualitative analysis of bibliographic data, we examined the foundational concepts of knowledge management provided by the most influential scholars and their most frequently cited publications.

References

1. Eom, S.B.: The Development of Decision Support Systems Research: A Bibliometrical Approach. The Edwin Mellen Press, Lewiston (2007). (3)
2. Turban, E., Sharda, R., Delen, D.: Decision Support and Business Intelligent Systems, 9th edn. Prentice Hall, Boston (2011). (150)
3. Eom, S.B.: Knowledge management and data mining: emerging business intelligence research subspecialties. In: Phillips-Ren, G., Carlsson, S., Respício, A., Brézillion, P. (eds.) DSS 2.0 - Supporting Decision Making with New Technologies, vol. 261, pp. 353–362. IOS Press, Amsterdam (2014)
4. Eom, S.B., Farris, R.: The contributions of organizational science to the development of decision support systems research subspecialties. J. Am. Soc. Inf. Sci. **47**, 941–952 (1996). (4)
5. Kruskal, J.B., Wish, M.: Multidimensional Scaling. Sage Publications, Newbury Park (1990)
6. Hair Jr., J.F., Anderson, R.E., Tatham, R.L.: Multivariate Data Analysis with Readings, 2nd edn. Macmillan Publishing Company, New York (1987)
7. Eom, S.B.: Author Cocitation Analysis: Quantitative Methods for Mapping the Intellectual Structure of an Academic Discipline. Information Science Reference, Hershey (2009)
8. Eom, S.: Mining Cocitation Data with SAS Enterprise Guide. Cambridge Scholars Publishing, New Castle upon Tyne (2015)
9. Eom, S.: Knowledge management and data mining: emerging business intelligence research subspecialties. In: Phillips-Wren, G.E., Carlsson, S., Respício, A., Brezillon, P. (eds.) DSS 2.0: Supporting Decsiion Making with New Technologies, vol. 261, pp. 353–362. IOS Press, Amsterdam (2014)
10. Argote, L., McEvily, B., Ray, R.: Managing knowledge in organizations: an integrative framework and review of emerging themes. Manage. Sci. **49**, 571–583 (2003)
11. Argote, L., Ingram, P.: Knowledge transfer: a basis for competitive advantage in forms. Organ. Behav. Hum. Decis. Process. **82**, 150–169 (2000)
12. Argote, L., Ingram, P., Levine, J.M., Moreland, R.L.: Knowledge transfer in organizations: leraning from the experiences of others. Organ. Behav. Hum. Decis. Process. **82**, 1–8 (2000)
13. Nonaka, I.: A dynamic theory of organizational knowledge creation. Organ. Sci. **5**, 14–37 (1994)
14. Nonaka, I., Takeuchi, H.: The Knowledge-Creating Company. Oxford University Press, New York (1995)
15. Davenport, T.H., DeLong, D.W., Beers, M.C.: Successful knowledge managemnet projects. Sloan Manage. Rev. **39**, 43–52 (1998)
16. Davenport, T.H., Prusak, L.: Working Knowledge: How Organizations Manage What They Know. Harvard Business Press, Cambridge (1998)
17. Alavi, M., Tiwana, A.: Knowledge integration in virtual teams: the potential role of KMS. J. Am. Soc. Inf. Sci. Technol. **53**, 1029–1037 (2002)

18. Alavi, M., Leidner, D.E.: Review: knowledge management systems: conceptual foundation and research issues. MIS Q. **25**, 107–136 (2001)
19. Bera, P., Burton-Jones, A., Wand, Y.: Guidelines for designing visual ontologies to support knowledge identification. MIS Q. **35**, 883–908 (2011)
20. Jukic, N., Nestorov, S.: Comprehensive data warehouse exploration with qualified association-rule mining. Decis. Support Syst. **42**, 859–878 (2006)
21. Chen, Y.-J.: Development of a method for ontology-based empirical knowledge representation and reasoning. Decis. Support Syst. **50**, 1–20 (2010)
22. Chen, G., Liu, H., Yu, L., Wei, Q., Zhang, X.: A new approach to classification based on association rule mining. Decis. Support Syst. **42**, 674–689 (2006)
23. Delen, D.: A comparative analysis of machine learning techniques for student retention management. Decis. Support Syst. **49**, 498–506 (2010)
24. Kuo, R.J., Lin, L.M.: Application of a hybrid of genetic algorithm and particle swarm optimization algorithm for order clustering. Decis. Support Syst. **49**, 451–462 (2010)
25. Chua, C.E.H., Storey, V.C., Chiang, R.H.L.: Deriving knowledge representation guidelines by analyzing knowledge engineer behavior. Decis. Support Syst. **54**, 304–315 (2012)
26. Wu, H., Gordon, M.D., Fan, W.: Collective taxonomizing: a collaborative approach to organizing document repositories. Decis. Support Syst. **50**, 292–303 (2010)
27. Choi, S.Y., Lee, H., Yoo, Y.: The impact of information technology and transactive memory systems on knowledge sharing, application, and team performance: a field study. MIS Q. **34**, 855–870 (2010)
28. Jackson, P., Klobas, J.: Transactive memory systems in organizations: implications for knowledge directories. Decis. Support Syst. **44**, 409–424 (2008)
29. Wang, J., Gwebu, K., Shanker, M., Troutt, M.D.: An application of agent-based simulation to knowledge sharing. Decis. Support Syst. **46**, 532–541 (2009)
30. Dong, S., Johar, M., Kumar, R.: Understanding key issues in designing and using knowledge flow networks: an optimization-based managerial benchmarking approach. Decis. Support Syst. **53**, 646–659 (2012)
31. Mohan, K., Jain, R., Ramesh, B.: Knowledge networking to support medical new product development. Decis. Support Syst. **43**, 1255–1273 (2007)
32. He, W., Wei, K.-K.: What drives continued knowledge sharing? an investigation of knowledge-contribution and -seeking beliefs. Decis. Support Syst. **46**, 826–838 (2009)
33. Eryarsoy, E., Koehler, G.J., Aytug, H.: Using domain-specific knowledge in generalization error bounds for support vector machine learning. Decis. Support Syst. **46**, 481–491 (2009)
34. Hsu, I.-C., Sabherwal, R.: Relationship between intellectual capital and knowledge management: an empirical investigation. Decis. Sci. **43**, 489–518 (2012)
35. Richardson, S.M., Courtney, J.F., Haynes, J.D.: Theoretical principles for knowledge management system design: application to pediatric bipolar disorder. Decis. Support Syst. **42**, 1321–1337 (2006)
36. Vapnik, V.: Statistical Learning Theory. Wiley, New York (1998)
37. Vapnik, V.: The Nature of Statistical Learning Theory. Springer, New York (2000)
38. Vapnik, V.N.: Estimation of Dependences Based on Empirical Data. Springer, New York (1982)
39. Quinlan, J.R.: Induction of decision trees. Mach. Learn. **1**, 81–106 (1986)
40. Quinlan, J.R.: Learning with continuous classes. In: Adams, N., Sterling, L. (eds.) Proceedings of the 5th Austrailian Joint Conference on Artificial Intelligence (AI 1992), vol. 92, pp. 343–348. World Scientific, Singapore (1992)
41. Quinlan, J.R.: C45: Programs for Machine Learning. Morgan Kaufmann, San Mateo (1993)

42. Baesens, B., Setiono, R., Mues, C., Vanthienen, J.: Using neural network rule extraction and decision tables for credit-risk evaluation. Manage. Sci. **49**, 312–329 (2003)
43. Hastie, T., Tibshirani, R., Friedman, J.: The Elements of Statistical Learning: Data Mining, Inference, and Prediction. Springer, New York (2001)
44. Witten, I.H., Frank, E.: Data Mining: Practical Machine Learning Tools and Techniques with Java Implementations. Morgan Kaufmann Publishers, San Francisco (2000)
45. Peng, Y., Kou, G., Shi, Y., Chen, Z.: A multi-crieria convex quadratic programming model for credit analysis. Decis. Support Syst. **2008**, 1016–1030 (2008)
46. Martens, D., Bruynseels, L., Baesens, B., Willekens, M., Vanthienen, J.: Predicting going concern opinion with data mining. Decis. Support Syst. **45**, 765–777 (2008)

Updating Business Intelligence and Analytics Maturity Models for New Developments

Louis Muller and Mike Hart$^{(\boxtimes)}$

Department of Information Systems,
University of Cape Town, Cape Town, South Africa
louisvell@gmail.com, mike.hart@uct.ac.za

Abstract. Recent developments such as real-time, social, predictive and cloud business intelligence and analytics (BI&A) introduce extra ways for organisations to obtain insight and business value from an expanded range of data. Organisations have struggled with the strategy, implementation, and measurement of their BI&A efforts, and a series of business intelligence maturity models (BIMMs) has been introduced to identify strengths and weaknesses of their BI&A situation, and assist remedial action. These BIMMs are however seen to be incomplete and outdated and do not accommodate recent BI&A developments. This study suggests how BIMMs should be modified to cater for these developments. Existing BIMMs were examined, and interviews conducted with BI&A professionals knowledgeable about BIMMs and recent BI&A changes. Findings suggested that existing BIMM dimensions should be modified in various ways to cater for the recent changes in BI&A. In addition, project management was identified as a new BIMM dimension.

Keywords: Business intelligence · Maturity models · Analytics · Big data · Decision support

1 Introduction

In 1958 an IBM researcher, HP Luhn, introduced the term "business intelligence", describing it (BI) as a decision making method based on understanding the interrelationships of different types of information so as to be able to make informed decisions and thereby reach organisational objectives [1]. Today business intelligence and analytics (BI&A) generally refers to the technology, applications and processes used to gather, store and analyse data, to assist people to make sound business decisions. For the past five years BI&A has been ranked a top priority for organisations globally as a tool for achieving a competitive advantage and attaining business value [2, 3].

While organisations embrace BI&A values by incorporating BI&A into their businesses, studies (e.g. [4, 5]) show that they have problems with measuring, implementing, and maintaining their current BI&A initiatives. Therefore, a series of BI maturity models (BIMMs) has been constructed that engage these problems on an organisational level. BIMMs measure and guide an organisation's BI&A efforts on a fixed series of ascending BI&A maturity levels, where each level is contained within a

© Springer International Publishing Switzerland 2016
S. Liu et al. (Eds.): ICDSST 2016, LNBIP 250, pp. 137–151, 2016.
DOI: 10.1007/978-3-319-32877-5_11

specified domain or dimension (e.g. technology, culture, people, infrastructure) that emphasises the strengths and weaknesses of an organisation's BI&A advances [6].

Most of these BIMMs are however viewed as outdated and irrelevant as they do not accommodate recent BI&A developments such as real-time, predictive, social and cloud BI&A. Authors in [4–7] argue that existing BIMMs cannot measure an organisation's ongoing BI&A progress accurately because of this. The aim of this research is to show how BIMMs can accommodate recent BI&A developments by identifying any new requirements or modifications that need to be considered to create a suitable revised maturity model. BIMMs consist of a series of dimensions that cover the entire value chain of an organisation's BI&A. This includes organisations focusing on sustainability and societal challenges. This research therefore aims to answer the following research questions:

Research Q1: *How should existing BIMM dimensions be modified to accommodate recent BI&A developments?*
Research Q2: *Which new BIMM dimensions should be taken into account to accommodate recent BI&A developments?*

This paper is organised as follows: Some background is given on the role of BIMMs in BI&A, and recent BI&A developments that may impact on them. The research methodology is then outlined, followed by an analysis of existing BIMMs. The research questions are answered through analysis of the interviewee responses, and the paper then concludes.

2   Background to the Research

2.1   BI&A Overview

Business Intelligence (BI) can be seen as an entire system with interrelated components working together to support informed business decisions. While [5, 8] explain that BI relates to decision support systems spread across technology, business processes, systems, and applications, [9] describe the systems and technology as the BI backbone consisting of IT components which support two main processes. The first process of production collects, cleans, transforms and stores source data as meaningful information. Business users then interrogate and analyse the data (process of consumption) in order to make informed decisions [10].

The term business analytics (BA) has recently become more fashionable. To some this defines only the often-neglected data mining, predictive or advanced analytics components of BI, while others take a far broader view. "For BA to actually work in an organization, there are issues quite aside from data management, number crunching, technology, systematic reasoning, and so on" [11, p. 139]. These authors mention factors such as awareness and commitment to the organization's vision, mission, and strategy; an analytics-friendly culture; and a management philosophy that understands and supports BA.

The use of BI&A as a combined term is suggested by [7, 12] for better understanding and simplification of research in the areas of BI and BA.

2.2 BI&A Maturity Models (BIMMs)

BI&A maturity models (BIMMs) were created to deal with problems related to measuring an organisation's BI&A effort. Organisations are continually experiencing problems in areas such as the implementation, measurement and maintenance of their current BI&A initiatives [4, 5]. The technological and organisational aspects of BI&A are a constant challenge in the implementation of BI&A [13].

BI maturity models (BIMMs) were proposed to address these problems holistically and to implement effective BI&A in organisations. The paper [6] explains that (a) BIMMs identify the strengths and weaknesses of an organisation's BI&A initiatives; and (b) BIMMs act as a roadmap to progress and plan an organisation's BI&A efforts to a more advanced state. Effective BI&A can be achieved only through analysing the overall BI&A design and structures holistically [14].

2.3 BIMM Characteristics

BIMMs consist of various features which uniquely identify their role in assessing BI&A maturity. According to [15, p. 219], a BIMM "consists of a sequence of maturity levels for a class of objects and thus represents an anticipated, desired, or typical evolution path of these objects shaped as discrete stages". The following five BIMM characteristics as described by [16] are used to measure an organisation's BI&A efforts:

- Object of Maturity Assessment: Key areas that present the BI&A environment.
- Dimension: Uniquely defined capabilities of each object of maturity assessment.
- Levels: These present the fixed state in which each object of assessment is assessed per dimension.
- Maturity Principle: (a) Continuous - scoring BIMM characteristics at different levels across domains OR (b) Staged - verifying if all BIMM characteristics of one level are achieved before advancing to the next level.
- Assessment: Conducted using either a quantitative or a qualitative approach.

2.4 Some Existing BIMMs

Table 1 lists 15 existing BIMMs, many of which have their origin in practice and have little documentation on their derivation. Their focus varies and is often largely technical. In many of the ten BIMMs they observed [14, p. 3] note that "topics like efficiency, organizational structures, staff, and strategy are rarely addressed", and suggested that a comprehensive BIMM should address all relevant dimensions. The three most recent BIMMs, originating from consulting organisations, also include Big Data concepts and terminology.

2.5 Recent BI&A Developments

Recent developments affecting BI&A include big data warehousing concepts, varied source data, new skill sets, architecture and analytics. According to [34], social media

Table 1. Some existing BIMMs

	BIMM	Source	Origin
1	Watson et al. (2001)	[17]	Academia
2	TDWI (2004, 2009)	[18, 19]	Practice
3	LOBI (2005)	[20]	Academia
4	AMR v2 (2006)	[21]	Practice
5	Sen et al. (2006, 2012)	[22, 23]	Academia
6	BI Maturity Hierarchy (2007)	[24]	Academia
7	HP (2007, 2009)	[25, 26]	Practice
8	Gartner (2008)	[27]	Practice
9	BIDM (2010)	[28]	Academia
10	EBIM (2011)	[29]	Academia
11	SOBIMM (2011)	[30]	Academia
12	EBI2 M (2012)	[4]	Academia
13	IDC – Big Data & Analytics MM (2013)	[31]	Practice
14	TDWI Big Data MM (2014)	[32]	Practice
15	IBM Big Data MM (2014)	[33]	Practice

and sensor device data have generated new technologies such as Hadoop, and increased demand for predictive analytics has resulted in the new area of Data Science. Cloud BI&A, real-time BI&A, social media analytics and predictive analytics are the recent developments having the most impact on BI&A [34–36].

Cloud BI&A. In recent years, cloud computing has become a major technological trend in IT. Cloud computing provides organisations the flexibility to expand current IT infrastructure and performance to cope with large processing and performance needs based on a pay for service and resource model [37]. Cloud computing scalability facilitates on-demand analysis of big data and real-time BI [38, 39].

Real-time or Near Real-time BI&A. Real-time or near real-time BI&A can give business users the ability to make on-demand decisions. According to [40], real-time BI&A presents freshly updated information ready for business users to make urgent decisions effectively. BI is extended to operational functions and threats and opportunities can be identified early on [40, 41].

Social Media Analytics. Social media data is now a huge addition to existing structured data, and a major component of big data. Social analytics has now gained wide use [12].

Predictive Analytics. Data mining has for decades been a less-used component of BI, but has lately been renamed as advanced analytics and predictive analytics, and is now increasing in importance and application [35, 36, 42].

The Four Vs of Big Data. There are numerous definitions of Big Data, but one of the most widely used encompasses the "Four Vs" of Variety, Volume, Velocity and Value. Variety acknowledges the vast increase in unstructured or semi-structured data such as social media, machine-generated data, images and video, all adding to the volume, and velocity describes both the increased frequency of generating data, and the desire to capture and use it more frequently. New approaches in supporting and accommodating BI&A are therefore required [37, 43].

3   Analysis of 15 Existing BIMMs

Published details of each of the 15 BIMMs selected were then analysed to generate a set of dimensions or domains represented in the set, and after some iteration 11 distinct dimensions emerged. Table 2 lists them, with an accompanying description.

Table 2. Emergent dimensions

Dimension	Description or example
BI&A applications and tools	e.g. OLAP; reporting; data mining
BI&A architecture	Structure of integration, sources, platforms
BI&A change management	Ability to manage changes
Data quality and use	Data quality, usage and management
Performance management	KPIs and metrics used to manage organisation
Skills and experience	BI&A competence and skill sets
IT infrastructure	Cloud computing, networks, servers, storage
Promoting BI&A culture	BI&A awareness and top management support
Knowledge management	Specific purpose is knowledge management
Business and IT alignment	Business and IT not in silos, enterprise BI&A
BI&A strategy	Managing overall BI&A strategy

Table 3 shows which of the emergent dimensions featured in each of the 15 BIMMs. The extent varied greatly, with Data Quality present in 13, while Knowledge Management (generally viewed as separate from BI&A) featured in only two and Change Management in only three of the 15 BIMMs.

3.1   Issues Identified from Existing BIMMs

Taking into account the summary in Table 3 and the recent developments of Sect. 2.5, the following issues emerged:

No BIMM Measurement Standard. The existing BIMMs all have their own criteria and focus of measurement. Some of the BIMMs have strong focus on the business aspects, while other BIMMs are concentrated on technical aspects, making it extremely difficult to select one BIMM overall for BI&A maturity assessment.

Low Representation of Some Dimensions. Only three out of the 15 BIMMs cover change management in BI&A. Only four BIMMs consider organisational performance management as a dimension. With new BI&A developments, much change is needed, and performance management could identify whether investment in these new BI&A areas was successful or not. Recent big data developments clearly require new skills in

Table 3. Dimensions in each of the BIMMs studied

Dimension	BI&A applications and tools	BI&A architecture	BI&A change management	Data quality and use	Performance management	Skills and experience	IT infrastructure	Promoting BI&A culture	Knowledge management	Business and IT alignment	BI&A strategy
Watson et al. (2001)	x	x	x	x	x	x					
TDWI (2004, 2009)	x	x		x			x	x			
LOBI (2005)	x			x			x		x		
AMR v2 (2006)				x	x			x		x	
Sen et al. (2006, 2012)		x	x	x			x	x			
BI Maturity Hierarchy (2007)				x			x		x		
HP (2007, 2009)	x					x	x			x	x
Gartner (2008)	x			x				x			x
BIDM (2010)		x		x	x						
EBIM (2011)	x	x		x						x	x
SOBIMM (2011)		x		x			x			x	x
EBI2M (2012)			x	x	x	x		x		x	x
IDC BigData&Analytics MM (2013)	x					x	x	x		x	x
TDWI Big Data MM (2014)		x		x		x	x	x			
IBM Big Data MM (2014)	x	x		x				x			x
Count of each Dimension	8	8	3	13	4	5	8	8	2	6	7

data science and predictive analytics, yet only five out of 15 BIMMs incorporate skills and expertise. (Knowledge management appears in only two BIMMs, but is generally considered to be separate from BI&A).

4   Research Methodology

Given the research subject and objectives, and widely differing published views on what should constitute a BIMM, an exploratory, interpretive and inductive approach was deemed most suitable, rather than using a positivistic philosophy. Interviewing of experts was thought to be the best way of gaining a deeper understanding of the issues surrounding the area. A purposeful sampling strategy is most widely used for this type of qualitative research [44], and a sample was selected of experienced BI&A practitioners with knowledge of both BIMMs and new developments in the IT area. Management and/or consulting experience and technical knowledge were required.

A semi-structured "conversational" interview prompt sheet was developed from material uncovered in the literature review and from the 15 BIMMs, consisting of open ended questions, so that interviewees could discuss their experiences with, and views of, the new developments and of BIMMs. Participants were first asked to discuss BIMMs in general. The interviewer then outlined the range of new developments that he suggested might be taken into account in future BIMMs. In order to ensure a wide exploration, he then initiated discussions in three separate broad areas: People, skills and experience; BI&A requirements, architecture and IT infrastructure; and BI&A culture, strategy, change and performance management. These focused on how the new developments might affect the BI maturity of organisations in these general areas. There were then a few wrap-up questions.

Interviews averaged about one hour each, allowing for follow-up of responses and for participants to introduce new issues. All interviews were recorded and transcribed, and thematic analysis was then used iteratively to develop themes, following the six phase process in [45, p. 87] of: data acquaintance; systematic coding; theme identification; theme reviewing; classifying and labelling themes; and generating the end report. After ten interviews the level of information gathered appeared to be close to saturation [46], and the final themes and subthemes emerged.

4.1 Participant Details

Table 4 shows the participants' current industry, role, and years of experience. For anonymity they will be referred to as P1, P2,..., P10.

Evidence of their experience is shown by comments such as:

"I used Gartner's maturity model and I have just built my own one." [P2].
"...be very selective of which parts of TDWI's model or Gartner's model is applicable to your organisation and where you want to be..." [P4].
"So practically in just my experience I used the BI maturity model to define that roadmap and also to sub-categorise the roadmap." [P6].

Table 4. Summary details of participants interviewed

Industry	No.	Industry experience	No.	Role	No.	Awareness of BIMMs	No.	Experience of BIMMs	No.
Retail	3	10–19 yrs	5	Head of BI	1	Not Aware	0	None	1
Finance	2	20–24 yrs	3	Exec Dir BI&A	1	Aware	1	Limited	3
Marketing	1	25+ yrs	2	BI Strategist	4	Strong Knowledge	9	Much	6
Consulting	4			BI&A Consultant	2				
				CIO	2				
Total	10		10		10		10		10

5 Analysis of Participant Interviews

This section covers points obtained from the thematic analysis of the participant interviews, relating to BIMMs in general, and their comments on how they could be updated for present day BI&A requirements. When participants were asked about BIMMs in general, they mentioned both benefits and concerns. The benefits were (a) the ability of a BIMM to act as a roadmap and (b) identifying the strengths and weaknesses of BI&A. This agrees with the benefits stated by [6].

> "...about having a mechanism to um...take a journey towards maturity and being able to firstly understand where that journey needs to go." [P9].
> "... identifying the strengths and weakness, that's our starting point and that kind of guides us." [P7].

There were three main concerns: Lack of a standard or focus is expressed by P4:

> "There's such a variety of models. They all measure different perspectives but there's not one that gives you a comprehensive view. There's no standard in that kind of model. Some of them lean towards only measuring the technology and the technical capabilities; others learn towards only measuring the management and the people and the processes about BI, but there's no one that gives you a balanced view on everything."

The view that a BIMM tended to be limited to a particular organisation type came from P6:

> "each organisation is unique so how they apply the analytics is defined by the specific organisation, the industry, the culture within the organisation."   ·

P9 stated that BIMMs were outdated, not accommodating recent developments:

> "Their [BIMMs] are still looking at BI that's you know kind of like the 1990 s and 2000 s, traditional BI. ...more recent technologies, things like descriptive analytics, real-time BI, things like big data analytics... things like data lakes, the maturity models don't speak particularly well to those."

Table 5 displays the themes and subthemes that emerged from the thematic analysis. Themes correlate fairly closely to the dimensions obtained from the literature analysis. Issues important for accommodating the recent developments are indicated as subthemes. Each dimension will now be discussed. Note that due to space limitations

Table 5. BIMM requirements to accommodate recent BI&A developments

BIMM dimensions	Sub-themes
Skills and experience	Leadership and management Hiring and identifying the right people Skills: Soft skills, Technical skills; Data science
BI&A architecture & IT Infrastructure	Accommodate Big Data 3 Vs (velocity, variety, volume)
BI&A and business alignment	BI business analysis skills Identifying requirements Strong BI&A partnership

(Continued)

Table 5. (*Continued*)

BIMM dimensions	Sub-themes
Performance management	"Near real-time" KPI measurement & drilldown
BI&A strategy	Research and development (R&D) strategy
BI&A culture	Top management support Promote value delivered by BI&A
Data quality and use	Data quality architecture or strategy
BI&A change management	Organisational support Communication plan/strategy Skills development and training.
Project management	Requirement for project management Adopting Agile project management

only the main changes due to new BI&A developments will be covered, and subthemes continuing as before will not generally be mentioned.

5.1 Skills and Experience

The importance of strong leadership and mentorship and an ability to embrace change brought by new developments is illustrated by the following quotes:

> "*I think the leadership needs to be very strong; the mentorship potential needs to be very high.*" [P3].
> "*It's a vital critical role and it can be very injurious to a company to have close minded people in leadership. Someone who supports new thinking, new ideas, and creative open minded; that's what you need*" [P5].

Participants underlined the importance of the BI&A leader selecting the right people, with suitable qualifications and experience, to support their future BI&A objectives. Three sets of skills were deemed important to meet the new BI&A requirements. Soft skills such as adaptability, open-mindedness and willingness to embrace change were often mentioned.

> "*So it's not just the skill it's also the willingness to do things differently and to challenge our own paradigms and from a process point of view it does also challenge the way we have done IT in general.*" [P1].

On the technical side, data integration and new programming skills were needed in the era of the 4Vs.

> "*... development skills, you still need coding skills, there's technical based capabilities that you will always need. ... if you are doing big data you have to have some kind of data integration capability.*" [P7].

Data Science skills such as machine learning, predictive analytics, mathematics and statistical modelling were also required:

> "*So for analytical skills, particularly the modelling skills, you need some type of statistical and synthesis background; the data scientist*" [P2].

5.2   BI&A Architecture and IT Infrastructure

The three previous dimensions of BI&A architecture, BI infrastructure, and BI&A applications and tools, have essentially been replaced by this single dimension. This is because of the potentially radical changes brought about by the "three Vs".

> *"you need some type of technology that can manage big data volumes and particularly with your text mining or ...sentiment analysis, you need to be able to ... clean it up and engage with it quickly."* [P2].

A scalable cloud solution was generally seen as being useful for doing so.

> *"So you're then looking at infrastructure [cloud BI&A] that allows you to move large volumes of data around the organisation."* [P9].

5.3   BI&A and Business Alignment

BI&A and business alignment emerged as a primary theme consisting of the following sub-themes: (a) BI Business analysis (BIBA) skills (b) identifying new requirements; (c) strong BI&A and business partnership. Participants specified that new requirements for recent BI&A developments should be identified. They suggested that BIBA skills are essential when it comes to identifying the new BI&A and business requirements and creating a business case. P2 noted:

> *"they enable technical guys but they also take the technical speak and they take it back to the business in layman's terms that they can understand."*

In commenting on BI being aligned with the new business requirements P4 said:

> *"...understanding your business drivers will allow you to develop a BI architecture and a BI model um...that will actually support your business."*

P1 pointed out the importance of a strong partnership between BI&A and business:

> *"If you want success for BI and want to increase the maturity of BI in your organisation it is a partnership between IT and business, it's co-creation."*

5.4   Organisational Performance Management

Corporate performance management functionality has often been included in BI reviews, hype cycles, etc. With real-time and near-real-time BI, measurements are taken more frequently, giving the opportunity for KPIs and analysis at a finer granular level time-wise than before. This would add to the existing important measurement, monitoring and drilldown features of organisational performance management, e.g.:

> *"With being able to measure more things accurately, you can set up hourly KPIs"* [P3].

5.5   BI&A Strategy

Two strategies were mentioned that have already been covered: being more predictive; and aligning BI&A with the business. A new area was a Research and Development (R&D) Strategy. Given the new data sources, technologies and methods of analysis, many participants stressed the importance of research. As P5 stated:

> "... new developments, they're experimenting with new technologies and new methodologies; they're trying to introduce disruptions into their world."

P8 discussed the need to devote time and space to R&D and learning on the job:

> "... 80/20 policy. Take at least twenty percent of your time and devote it to just learning what is going on in the field at the moment, and have a pitch back to the business every month or three months."

5.6   BI&A Culture

The necessity of top management support for catering for the new BI&A directions (preferably from the CEO), and BI champions was greatly emphasised:

> "I think the best thing is that you need buy in on an executive level. So your BI champion needs to be at an executive level as well." [P2].
> "It's top down, the CEO starts asking those questions continuously to his ten directors who bought into him. They're going to start asking it to those that report and that's the easiest way to establish a new culture..." [P7].

Top management support can be a useful enabler for the second sub-theme of promoting the value that new BI&A developments can deliver, but as P2 points out:

> "a big challenge at the moment is not all companies understand the value of BI and how you can use data and with analytics it's a huge opportunity."

5.7   Data Quality

Table 4 showed data quality to be the most consistent dimension across existing BIMMs. Although it was not mentioned at all in the prompt sheet questions, the importance of data quality was reinforced by interviewees, with strong calls for data quality architecture and governance appropriate to the new requirements generated by the three Vs of variety, velocity and volume. For example:

> "Data quality is core. You have to have a technical capability to address data quality and data conformity you need data quality management architecture." [P7].

5.8   Change Management

Only three of the 15 BIMMs examined included change management, yet participants viewed this as extremely important given changing needs for architecture, skills, experience and general BI&A approaches. There were three sub-themes: (a) skills development, (b) organisational support and (c) communication plan or strategy.

"whatever language you were using five years ago, I guarantee it's not being used anymore because the world has moved on.." [P8].
"you need to perform change management in your organisation in terms of BI...the most important thing in change management will be people." [P4].

Participants felt that the whole organisation should be included as part of the change management process.

"...if you want someone to buy into it you need to include them. So on the journey I used to have whenever we had like a new session I used to include one person from every business unit at my start session." [P2]. P10 noted:
"you are not just introducing new technologies you introducing new ways of accessing information which fundamentally can change business processes."

Communication with external stakeholders such as vendors, consultants and clients, as well as throughout the organisation, was essential.

"Communication strategy is also incredibly important...goes hand in hand with change management. So you need an internal communication strategy as well as an external communications strategy." [P6].

5.9   Project Management – A New Dimension

Project management was not included in any of the 15 BIMMs studied, but was mentioned as important by several participants, as a project management plan with milestones and deliverables improves ability to cater for recent BI&A developments.

"to mark milestones in the roadmap.. a roadmap is very important but I think milestones are even more important so that you can get to a point." [P3].

Interviewees strongly recommended adopting agile project management methodology because BI&A and business requirements are constantly changing and the organisation's BI&A strategy must remain relevant and satisfy business goals. Project benefit and value realisation should also be analysed and followed up.

5.10   Summary

Analysis of the interviews produced nine themes very similar to the dimensions obtained from the review of the existing 15 BIMMs. The main difference was that the previous three separate dimensions of BI&A architecture, IT infrastructure, and BI&A applications and tools have combined to form one new dimension of BI&A architecture and IT infrastructure. This is due to the major changes in approach that may be brought about by storage and analysis of "three V's" data, and use of the cloud. Due to the extent of these changes, a new important dimension of project management emerged. The previous sections have highlighted the important subthemes that organisations may need to consider within each BIMM dimension.

Potential limitations are that the interviewees, although all experienced BI&A practitioners, may have limited exposure to massive varied datasets incorporating all 3 Vs. Thematic and interpretive analysis carried out by another researcher might yield slightly different results. In addition, space limitations of this paper mean that it can not describe the further research that was carried out into what should be done in each dimension to achieve BI&A maturity.

6 Conclusion and Recommendations

Appropriate BIMMs can be very useful in facilitating successful implementation, maturity or pervasiveness of DSS, BI or business analytics across a wide range of organisations, including those focusing on "sustainability and societal challenges". This research has shown that for organisations aiming to achieve BI&A maturity in the new era of big data analytics and the cloud, existing BIMMs need to be updated and broadened.

A new BIMM dimension is that of Project Management, to effectively deal with the many potential changes arising from the new developments in IT and BI&A.

Other dimensions need to change or expand their scope. New IT and BI&A architecture, many new skills, experience and R&D strategy are required; top management needs to appreciate and support a culture of data-driven decision-making throughout the organization, and increased change management skills are needed to facilitate this and improve alignment between BI&A and business goals. It is unlikely that a single "one-size-fits-all" BIMM is feasible, and future research should examine contingency models suitable for different organisational BI&A situations. It could for example focus on a specific product or service-related industry.

Acknowledgements. This work is based on research partly supported by the South African National Research Foundation.

References

1. Luhn, H.P.: A business intelligence system. IBM J. Res. Dev. **2**(4), 314–319 (1958)
2. Gartner: Flipping to digital leadership: insights from the 2015 Gartner CIO agenda report (2015). www.gartner.com/cioagenda
3. Kappelman, L., McLean, E., Johnson, V., Gerhart, N.: The 2014 SIM IT key issues and trends study. MIS Q. Executive **13**(4), 237–263 (2014)
4. Chuah, M., Wong, K.: Construct an enterprise business intelligence maturity model (EBI2 M) using an integration approach: a conceptual framework. In: Mircea, M. (ed.) Business Intelligence-Solution for Business Development, pp. 1–12. InTech, Rijeka (2012)
5. Raber, D., Wortmann, F., Winter, R.: Towards the measurement of business. In: Proceedings of the 21st European Conference on Information Systems, pp. 1–12 (2013)
6. Dinter, B.: The maturing of a business intelligence maturity model. In: Proceedings of the AMCIS 2012, Paper 37 (2012)
7. Chen, H., Chiang, R.H., Storey, V.C.: Business intelligence and analytics: from big data to big impact. MIS Q. **36**(4), 1165–1188 (2012)

8. Elena, C.: Business intelligence. J. Knowl. Manage. Econ. Inf. Technol. **1**, 101–112 (2011)
9. Ponelis, S.R., Britz, J.J.: A descriptive framework of business intelligence derived from definitions by academics, practitioners and vendors. Mousaion **30**(1), 103–119 (2013)
10. Peter, W., Maria, C., Ossimitz, L.: The impact of business intelligence tools on performance: a user satisfaction paradox? Int. J. Econ. Sci. Appl. Res. **3**, 7–32 (2012)
11. Holsapple, C., Lee-post, A., Pakath, R.: A unified foundation for business analytics. Decis. Support Syst. **64**, 130–141 (2014)
12. Lim, E.P., Chen, H., Chen, G.: Business intelligence and analytics: research directions. ACM TMIS **3**(4), 17:1–17:10 (2013)
13. Luftman, J.N., Ben-Zvi, T.: Key issues for IT executives 2009: difficult economy's impact on IT. MIS Q. Executive **9**(1), 49–59 (2010)
14. Lahrmann, G., Marx, F., Winter, R., Wortmann, F.: Business intelligence maturity models: an overview. In: 44th HICSS, pp. 1–10 (2011)
15. Becker, J., Knackstedt, R., Pöppelbuß, J.: Developing maturity models for IT management - a procedure model and its application. Bus. Inf. Syst. Eng. **1**(3), 213–222 (2009)
16. Raber, D., Winter, R., Wortmann, F.: Using quantitative analyses to construct a capability maturity model for business intelligence. In: 45th HICSS, pp. 4219–4228 (2012)
17. Watson, H.J., Ariyachandra, T.R., Matyska Jr., R.J.: Data warehousing stages of growth. Inf. Syst. Manage. **18**(3), 42–50 (2001)
18. Eckerson, W.W.: Gauge your data warehouse maturity. DM Rev. **14**(11), 34 (2004)
19. Eckerson, W.W.: TDWI's Business Intelligence Maturity Model. TDWI, Chatsworth (2009)
20. Cates, J.E., Gill, S.S., Zeituny, N.: The ladder of business intelligence (LOBI): a framework for enterprise IT planning and architecture. Int. J. Bus. Inf. Syst. **1**(1), 220–238 (2005)
21. Hagerty, J.: AMR research's BI / performance management maturity model, Version 2 (2006). http://www.cognos.com/pdfs/analystreports/ar_amr_researchs_bi_perf.pdf
22. Sen, A., Sinha, A.P., Ramamurthy, K.: Data warehousing process maturity: an exploratory study of factors influencing user perceptions. IEEE Trans. Eng. Manage. **53**(3), 440–455 (2006)
23. Sen, A., Ramamurthy, K., Sinha, A.P.: A model of data warehousing process maturity. IEEE Trans. Softw. Eng. **38**(2), 336–353 (2012)
24. Deng, R.: Business intelligence maturity hierarchy: a new perspective from knowledge management. Inf. Manage., 23 March 2007 (2007)
25. Henschen, D.: HP touts neoview win, banking solution, BI maturity model. Intell. Enterp. **10**(10), 9 (2007)
26. Packard, H.: The HP Business Intelligence Maturity Model: Describing the BI Journey. Hewlett-Packard Development Company L.P. (2009)
27. Rayner, N., et al.: Maturity Model Overview for Business Intelligence and Performance Management. Gartner Inc. Research, Stamford (2008). http://www.gartner.com
28. Sacu, C., Spruit, M.: BIDM - the business intelligence development model. In: 12th International Conference on Enterprise Information Systems, SciTePress (2010)
29. Tan, C.-S., Sim, Y.-W., Yeoh, W.: A maturity model of enterprise business intelligence. Commun. IBIMA **2011**, 1–9 (2011)
30. Shaaban, E., Helmy, Y., Khedr, A., Nasr, M.: Business intelligence maturity models: toward new integrated model. Nauss.Edu.Sa. (2011). http://www.nauss.edu.sa/acit/PDFs/f2873.pdf
31. Vesset, D., Versace, M., Girard, G., O'Brien, A., Burghard, C., Feblowitz, J., Osswald, D., Ellis, S.: IDC MaturityScape: Big Data and Analytics - A Guide to Unlocking Information Assets, 20 p. IDC (2013)
32. Halper, F., Krishnan, K.: TDWI Big Data Maturity Model Guide. TDWI (2014)
33. Nott, C.: Big Data & Analytics Maturity Model (2014). http://www.ibmbigdatahub.com/blog/big-data-analytics-maturity-model

34. Wixom, B., Ariyachandra, T., Goul, M., Gray, P., Kulkarni, U., Phillips-Wren, G.: The current state of business intelligence in academia. CAIS **29**, 299–312 (2014)
35. Aziz, M.Y.: Business intelligence trends and challenges. In: BUSTECH 2014, the 4th International Conference on Business Intelligence and Technology, pp. 1–7. IARIA (2014)
36. Doshi, A., Deshpande, A.V.: Comparison analytics: future trends and implications. Int. J. Curr. Eng. Technol. **4**(5), 3253–3256 (2014)
37. Assunção, M.D., Calheiros, R.N., Bianchi, S., Netto, M.A.S., Buyya, R.: Big data computing and clouds: trends and future directions. JPDC **79–80**, 3–15 (2015)
38. Hashem, I.A., Yaqoob, I., Badrul Anuar, N., Mokhtar, S., Gani, A., Ullah Khan, S.: The rise of big data on cloud computing: review and open research issues. Inf. Syst. **47**, 98–115 (2014)
39. Chang, V.: The business intelligence as a service in the cloud. Future Gener. Comput. Syst. **37**, 512–534 (2014)
40. Tank, D.: Enable better and timelier decision-making using real-time business intelligence system. Int. J. Inf. Eng. Electron. Bus. **7**(1), 43–48 (2015)
41. Shi, Q., Abdel-Aty, M.: Big data applications in real-time traffic operation and safety monitoring and improvement on urban expressways. Transp. Res. Part C Emerg. Technol. **1**, 1–15 (2015)
42. Laxmi, P.S.S., Pranathi, P.S.: Impact of big data analytics on business intelligence-scope of predictive analytics. Int. J. Curr. Eng. Technol. **5**(2), 856–860 (2015)
43. Chen, M., Mao, S., Liu, Y.: Big data: a survey. Mob. Netw. Appl. **19**(2), 171–209 (2014)
44. Creswell, J.W.: Qualitative Inquiry and Research Design: Choosing among Five Traditions. Sage, Thousand Oaks (1998)
45. Braun, V., Clarke, V.: Using thematic analysis in psychology. Qual. Res. Psychol. **3**(2), 77–101 (2006)
46. Strauss, A., Corbin, J.: Basics of Qualitative Research: Techniques and Procedures for Developing Grounded Theory, 2nd edn. Sage, Newbury Park (1998)

A Novel Collaborative Approach for Business Rules Consistency Management

Nawal Sad Houari[✉] and Noria Taghezout

LIO Laboratory, Department of Computer Science,
University of Oran1 Ahmed BenBella, Oran, Algeria
sad.houari.nawal@gmail.com, taghezout.nora@gmail.com

Abstract. This paper presents an approach based on ontology and agents. The major objective is to automatically manage the consistency of business rules introduced by the experts during the capitalization of business rules process as part of a collaborative system dedicated to experts. The Evaluator agent is at the heart of our functional architecture, its role is to detect the problems that may arise in the consistency management module and provide a solution to these problems in order to validate the accuracy of business rules. It uses the knowledge represented in the domain ontology. We exploit the possibilities of TERMINAE method to represent the company's business model and manage the consistency of the rules that are introduced by business experts. The suggested approach treats here the cases of contradiction, redundancy, invalid rules, the domain violation and the rules never applicable. We conducted some experiments to test the feasibility of our approach.

Keywords: Business rules (BR) · Business rules management system (BRMS) · Collaboration · Consistency management · Multi-agents systems (MAS) · Ontologies

1 Introduction

In today's changing world, the success of a business depends heavily on its ability to quickly adapt itself to its market and its competition. As more and more business processes are being automated, there is a growing need for Information Technologies (IT) that can cope with change.

The big problem today is that the system knowledge is always embedded within the minds of business experts, and that knowledge is considered as capital of the enterprise. However, enterprises can survive after the departure of those experts that manage the business processes; the solution is the use of business rules modeling; this will encapsulate that knowledge as a content that can be managed by the company, in a format that allows an easy transition during personnel turnover. Another advantage of business rules modeling is harvested when the business knowledge expressed as business rules can be analyzed automatically, yielding an opportunity to infer additional business intelligence embedded within the set of rules.

© Springer International Publishing Switzerland 2016
S. Liu et al. (Eds.): ICDSST 2016, LNBIP 250, pp. 152–164, 2016.
DOI: 10.1007/978-3-319-32877-5_12

Sometimes the business rule is integrated into a database table structure, while at other times a business rule is embedded in an application program code or carried in the brains of business people [1]. In this paper, we focus on the business rules carried in the brains of business experts.

In fact, in computing science, a business rule is a high-level description that allows controlling and/or taking a decision, using enterprise specific concepts. Thus, the business rules describe what an expert needs to do to take a decision [2]. They capitalize the enterprise knowledge and translate its strategy by describing the actions to lead for a given process. They are generally written in a natural controlled language [3]. The rules can be defined in the form of simple rules (as IF <conditions> THEN <Actions>), decision tables or decision trees. In other words, a business rule is a directive intended to command and influence or guide business behavior, in support of business policy that is formulated in response to an opportunity or threat. From the information system perspective, a business rule is a statement that defines or constrains some aspect of the business. It is intended to assert business structure or to control or influence the behavior of the business [1]. Halle defines the business rules approach to systems development as that which "allows the business to automate its own intelligent logic better, as well as to introduce change from within itself and learn better and faster to reach its goals" [4]. Specifically, collaborative systems enhance the communication-related activities of a team of individuals that is engaged in coordination activities such as computer-assisted communication and problem solving, and help in the evaluation of a decision process.

The article is organized as follows; in the next section, we describe some related works and our contribution. In Sect. 3, we present our approach in details. To illustrate the feasibility of our proposed approach, the experimentations results are given in Sect. 4. Finally, Sect. 5 concludes our study and out-lines our future research directions.

2   Related Works and Contribution

The work presented in [5] tackles the rule acquisition problem, which is critical for the development of BRMS. The proposed approach assumes that regulations written in natural language (NL) are an important source of knowledge but that turning them into formal statements is a complex task that cannot be fully automated. Authors propose to decompose the acquisition process into two main phases: the translation of NL statements into controlled language (CL) and their formalization into an operational rule base. As a limitation of this work, the rules acquisition from natural language is a complex task and cannot enumerate all the possible cases. The research reported in [6] is aimed at designing a dynamically adaptable data-sourcing service for deploying business rules effectively in supply chain management. Authors propose a semantics-based approach to implement the data-sourcing service for business rules. The approach captures semantics of business rules and provides an agent-enabled mechanism that dynamically maps business rules to the enterprise data model. In other words, it bridges the semantic gap between business rules and the enterprise data model by mapping business terms in business rules to data objects in the enterprise data model. The rules and procedures are often provided in text, the authors in [7] propose a method and tools that enable the business rules acquisition from texts. Thus, they build and operate a documented rules model, a structure of "index" which connects the source text, the ontology that defines the conceptual vocabulary of the domain and the rules that are drawn from the text and are expressed with terms conceptualized in the ontology. The paper in [8] describes a simple formalism designed to encode lexicalized ontologies and shows how it is used in a business rule management platform of the automotive domain. Authors propose building an ontology as a formal model for representing conceptual vocabulary that is used to express business rules in written policies. OWL-DL (Ontology Web Language Description Logics) language is used to represent concepts and properties of the domain ontology but such an ontology must be linked to the lexicon used to express rules in the text, so experts can query source documents. This calls for a formalism to link linguistic elements to conceptual ones. Authors opt to use the SKOS[1] language which provides basic elements to link domain concepts to terms from the text. The combination of OWL entities, SKOS concepts and their related information form a lexicalized ontology which supports the semantic annotation of documents. In [9, 16], authors show two prototypes based on the BRMS WODM[2], (1) The OWL plug-in and (2) the change-management plug-in. The OWL plug-in enables authoring and executing business rules over OWL ontologies. It consists of importing OWL ontologies into WODM and using all the functionalities offered by this BRMS to author and execute rules. The change-management plug-in enables the evolution of business rules with respect to the ontology changes. This component, implemented basically using an OWL ontology and rules, detects inconsistencies that could be caused by an ontology evolution and proposes solution(s), called repair, to resolve them. The aim of the paper presented in [10] was to indicate possible applications of rule-based approach in production planning and scheduling. Many solutions described in the study can be implemented also in the computer decision support systems for cast iron manufacturers.

2.1 Our Contribution

Our paper aims to transfer knowledge from SMEs' experts to formal representations, which allow systems to reason with such knowledge. The main idea behind this study is to join the agent-based modeling and ontology-based approach, in order to take benefit of the advantages of the two.

The major objective is to automatically manage the consistency of business rules introduced by the experts during the capitalization of the business rules process as part of a collaborative system. Our proposal is mainly based on agents such as Expert agent, Evaluator agent, Translator agent, Supervisor agent and Security agent.

Our work is dedicated to the knowledge capitalization of domain experts in maintenance field. The knowledge management process involves two phases: first, the experts have an editor for entering their knowledge, all the update processing, and safeguard are insured. The consistency management of knowledge (rules) is also provided, and the case of redundancy, contradiction, invalid rules, domain violation and rules ever

---

[1] Simple Knowledge Organization System.
[2] WebSphere Operational Decision Management.

applicable are investigated and verified by the system. Second, in the case of incoherent rule, a collaboration session is initiated between experts; this step can lead to a negotiation among participants. This model enables effective collaborative decision-making and also facilitates exchanges and knowledge sharing between the different actors in safety.

We can summarize our contribution in the following key-points:

- Knowledge acquisition from business experts through a user-friendly and ergonomic web editor.
- Design and development of domain ontology to generate the company's business model.
- Design and implementation of an agent-based architecture, where the Evaluator agent plays an important role.

3   Proposed Approach

Our approach permits to capitalize the business experts' knowledge as business rules by using an agent-based platform. This latter offers facility to the experts as standardization and auditability of the business rules. We propose to use domain ontology in order to generate the business model corresponding to the enterprise.

The business rules management system is composed of several components as described in Fig. 1. Each component is detailed in [11, 12].

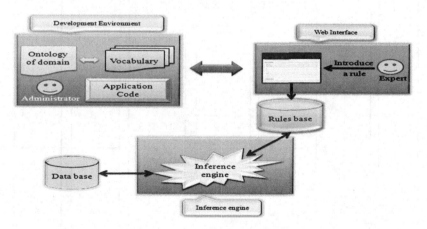

Fig. 1.   The proposed system [11]

3.1   Business Expert Knowledge Capitalization Process

Thanks to the generated business language, the business expert can write the rules in an autonomous manner. A rule is composed of a condition part and an action part; therefore the expert must specify the two parts. The process needs to pass through several steps, until the final storage in the rules base. We use a multi agents system that is composed of several agents:

1. Expert Agent: is responsible for the recuperation of the rules seized by the expert. This agent saves the rules and transmits them to Evaluator agent.
2. Supervisor Agent: performs all control tasks in the system.
3. Translator Agent: retrieves the rule from the Evaluator agent, and translate the introduced rule into technical rule.
4. Evaluator Agent: is responsible to assessing the consistency of the business rules. It recovers the rule from the Expert agent, browses the domain ontology to extracts the set of concepts that correspond to the introduced rule and accesses to the rules repository to test if this rule poses a problem with another rule, if it is the case then the Evaluator agent send a message to the Expert and Translator agents, otherwise it validates the rule.
5. Security Agent: is responsible to encrypting and decrypting the business rule [13, 14].

The interaction among the different agents of the system is shown by the sequence diagram (in AUML) which is represented in the Fig. 2.

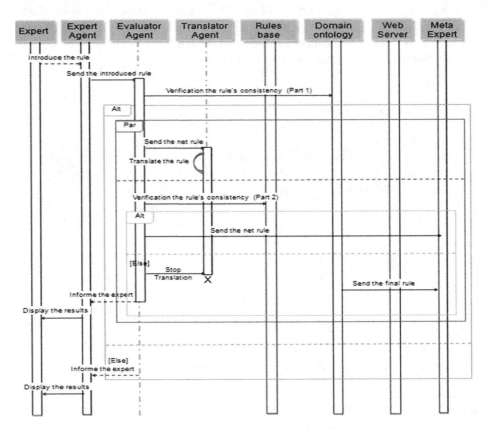

Fig. 2. Sequence diagram of our system

There are several possibilities with knowledge management in MAS:

- Every agent has knowledge about its problem domain.
- Each of the agents has its own knowledge representation.
- An agent only sends a message to all the agents which might be interested that some new information on certain subject exists.
- Every agent has knowledge about its problem domain, but whenever something new arises about the common knowledge which might be of interest for other agents, it send an inform message.

3.2   Business Rules Consistency Management

Currently, the effectiveness and correctness of business rules defined by the experts have always been a challenging problem. Our system should ensure that the complete set of business rules includes only the rules that are consistent and do not conflict among themselves. A possible way is to group the rules by the objects they constrain or actions they trigger and to check if there are any conflicts [15].

In what follows, we will present the different types of inconsistencies that may impact a set of rules, and the solution brought to each type of inconsistency.

1. **The contradiction:** a contradiction is detected in a set of rules R, if this set contains at least two rules that have the same condition part and assign two different values to a same attribute in the action part.

   If this type of problem is detected, the system sends notification to the expert who introduced this rule with a very detailed report on the problem description.

   The expert must respond to this notification and provide a solution to this problem as soon as possible, either by deleting or changing the rule.
2. **The rules never applicable:** a rule is never applicable if its condition part can never be verified.

   The system sends notification to the expert if this type of problem is detected with a very detailed description on the part of the rule is causing the problem as well as the range and interval of values requested for this attribute.

   The expert must respond to this notification in the briefest delays and modifies the rule according to the sent report. In the case where the expert wants to modify the values or the properties of an attribute, so he should contact the administrator to update the vocabulary of the domain according to new market changes.
3. **The domain violation:** a domain violation is detected in a set of rules, if this later contains at least one rule with a particular statement: it means, in the action part, the assigned value to the attribute is out of its domain.

   If this problem is detected, the business expert receives notification with recommendations on the editor with an explanation of the problem encountered as well as the values and properties allowed for this attribute. The expert must answer to this notification in the briefest delays and modifies the rule according to the sent report.
4. **The invalid rules:** a rule is invalid if it uses in its premise or conclusion part, a concept or a property that does not belong to the ontology from which the rules were

edited. To check if the rule is valid or not, the system accesses the domain ontology and follows these steps:

   a. Search if the concept exists, if the concept does not exist then the system sends a notification message to the business expert and delete the rule, otherwise go to b.

   b. Search if the property exists, if it is not the case then the system sends a notification to the business expert and delete the rule, otherwise the system assumes that the rule is valid.

5. **The redundancy:** two rules are called redundant if they have the same condition and the two rules produce the same action.

    If this type of problem is detected, the system deletes the rule and sends notifications to the expert to inform him that the rule creates a redundancy problem and has been deleted.

6. **The equivalence:** two rules are equivalent if the condition of the first one is included in the condition of the second and the both rules produce the same action. To solve this problem, the system sends a message containing the equivalent rules to the expert and asked him to integrate and merge all equivalent rules into a single coherent rule.

3.3 The Collaboration Between Experts in the Case of Inconsistencies

In the case of inconsistencies, the business expert receives a notification on the editor to correct its introduced rule according to the sent report and the erroneous rule will be stored in a temporary basis. If the expert doesn't answer before XX days, the system sends to him an e-mail and SMS to correct the rule. If the expert doesn't answer after XX days, therefore the system sends the rule as well as the assessment report to the other experts and to the meta-expert to ask for their help and their opinion.

Each expert can answer by «Yes» or «No». In the case of «Yes», the expert should send the new modified and corrected rule.

After harvesting responses from business experts, a negotiation process is launched to decide. If all experts choose «No» so the rule will be definitely deleted and the system sends a notification message to the expert who introduced this rule, otherwise the experts can correct the wrong rule and this later pass through the reparation process.

The negotiation mechanism among the experts is based on the well-known Contract Net Protocol. The contract Net protocol is a model for which only the manager emits propositions. The contractors can only make an offer but not counter-propositions. In our case, we propose to extend the CNP to consider the opinion of contractors (experts), in order to find more quickly a common accepted solution. The Contract Net Protocol is used because of its facility to implement negotiation protocols. We will detail our negotiation protocol in a future paper.

3.4 Building Domain Ontology

Any rule-based application usually starts in the development environment with collaboration between developers, business analysts and business experts. This step consists in defining the business model and the rules model as well as some functions necessary for the application development.

Domain experts who are not also business rules experts may have difficulties expressing their knowledge in formalized logic languages. Supporting them in their management of the knowledge needed to write these rules is one of our goals.

One of the difficulties with business rules is that various departments or roles sometimes use different vocabularies for the same things so they cannot understand each other immediately. In addition, formalized rules are often not easy to understand [8]. We propose building ontology as a formal model for representing conceptual vocabulary that is used to express business rules. Using a normalized vocabulary helps domain experts in writing rules more efficiently and is less costly than managing controlled vocabulary. Using an ontology as a unified model for a heterogeneous vocabulary and annotating the rules and the underlying ontology has another advantage: it will reduce misunderstandings and ensure that people are discussing the same thing.

Our domain ontology is developed by acquiring knowledge from documents, collection and capitalization of business rules processes with domain experts and the interviews with company managers. Currently, we have implemented our ontology in Protégé 4.0.2. Figure 3 depicts an overview of the main concepts and classes of the domain ontology that is applied to a maintenance enterprise.

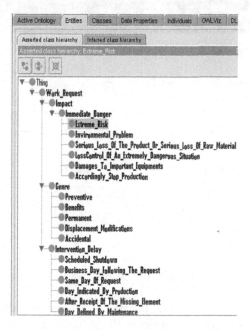

Fig. 3. A snapshot of our domain ontology

We have created our ontology according to the method described in the following:

1. Our ontology is built manually,
2. Ontology Enrichment: we use the TERMINAE method which is a method and platform that assists users in designing terminological and ontological resources from texts [17].

4   Issues of Implementation

This section is divided in two parts; the first one is dedicated to the collaborative graph-ical interface: here the experts will be able to introduce their knowledge in a simple form (rule). The editor will give them the possibility to update their rules, visualize the contents and launch simple or advanced research by using some keywords.

In the second part, we will present some experiments of business rules consistency management.

4.1   Execution Scenario

We consider here a simple example to illustrate our approach:

The expert accesses to the developed platform to capitalize his knowledge and his experience in a particular domain. Once the access is guaranteed, the business rule editor will be visualized as described in Fig. 4.

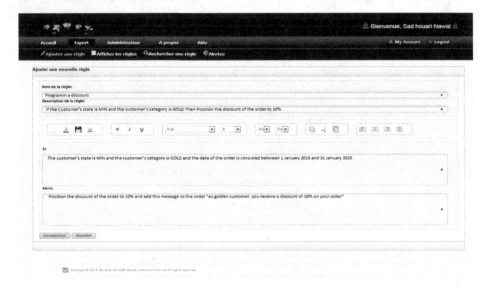

Fig. 4. Business rules editor

The expert can search for a rule, the developed platform offers to the business experts' an advanced research, based essentially on some keywords or criteria [11, 12]. The search result is shown in table form which contains all the information related to the business rule(s). By using the platform, the expert will be able also to update or delete the rule.

4.2   Experimentations

To study the behavior of our business rules consistency management approach, we designed and implemented a tool that defines the functioning of our approach. We will visualize, analyze and discuss the obtained results in the following.

4.2.1   Experiment 1: Test of Our Consistency Management Module

In order to test our system, we launched 60 business rules which contain: 12 contradictory rules, 8 redundant rules, 10 equivalent rules, 12 invalid rules, 11 rules never applicable and 7 rules that pose a domain violation problem (see Table 1). We note that our rule base already contains 30 consistent rules.

Table 1. Business rules set

Inconsistency type	Rules number	Number of rules detected
Invalid rules	12	12
Domain violation	7	7
Rules never applicable	11	11
Redundancy	8	8
Contradiction	12	12
Equivalence	10	0

The obtained results (see Fig. 5) are very encouraging and show that the developed system treats the problem of contradiction, redundancy, domain violation, invalid rules and rules never applicable but does not treat the equivalence problem equivalence.

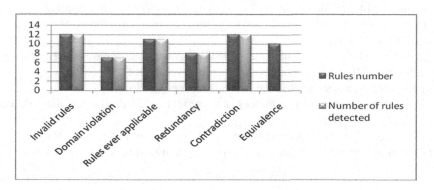

Fig. 5. Results of our consistency management system

4.2.2   Experiment 2: Evaluation of the Platform Usability

An evaluation of system usability using a questionnaire was conducted to show the effectiveness of the proposed platform.

We get inspired by the questionnaire given in [18] and we develop our questions regarding the system we suggested.

The questionnaire was distributed to 10 users consisting of experienced and un-experienced users, including business experts from the enterprise.

Figure 6 shows the usability results obtained from the questionnaire, in which a question is usually followed by a reversed question to reveal opposing facts. For each question, we assigned a weight. At the end we count the sum of weight to find the user's satisfaction level.

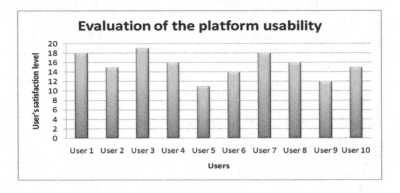

Fig. 6. Evaluation of the platform usability

Questionnaire content:

- If you have three words to describe the platform, what are these words?
- Which part of the platform did you find most interesting?
- Which part of the platform do you want to discuss or treat in more depth? And why?
- Would you like to add anything else? Do you have any other comments?
- On a scale of 0–5, how would do you value the performance of the platform (response time and access)?
- On a scale of 0 to 5, how much you value the ergonomics/the ease of use of the tool?
- On a scale of 0 to 5, how much you value the overall tool?
- There were inconsistencies in the navigation? Yes Or No
- The interaction components (buttons, menus, text fields, drop-down lists, etc.) can be easily understood? Yes Or No
- Do you need to learn many things before using the platform? Yes Or No
- Is it easy and effective to share your experiences in the platform? Yes Or No

From the questionnaire results, one could argue that the usability/readability of the platform is reasonably high.

4.2.3   Experiment 3: Agents Performance

The Fig. 7 presents the execution time of the system agents in milliseconds. We note that the execution time of the Expert agent is equal to 281 ms, the execution time of the Translator agent is equal to 580 ms and the execution time of the Evaluator agent is equal to 720 ms. The execution time of the Evaluator agent depends greatly on the ontology size.

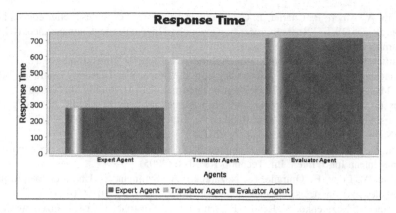

Fig. 7. Response time of the system agents

5 Conclusion

This paper proposes a novel collaborative approach for business rules consistency management in a typical small and medium enterprise. The suggested approach is utilizing domain ontology within an agent- based architecture.

The modeling is based on agents to increase the execution speed of processes and effective response. We propose building an ontology as a formal model for representing conceptual vocabulary that is used to express business rules. Such an ontology not only gives the vocabulary to be used in expressing the rules, it also provides a structured vocabulary that encodes relationships between concepts and supports checking for inconsistencies.

Our collaborative system provides:

- An editor easy to use,
- It offers features that stimulate business users and developers to use it,
- Automate precise, highly variable decisions,
- Easy, safe and predictable rule management.

In future works, we envisage to test and evaluate our prototype in other SME enterprise with a large number of rules, and handle the problem of equivalent rules.

For the executions of the rules: the RÊTE algorithm is intended to be used to reach this goal.

References

1. Business rules. In: Business Intelligence, chap. 11. http://dx.doi.org/10.1016/B978-0-12-385889-4.00011-9
2. Legendre, V., Petitjean, G., Lepatre, T.: Gestion des règles «métier». Softw. Eng. (92), 43–52 (2010)

3. Chniti, A., Albert, P., Charlet, J.: Gestion de la cohérence des règles métier éditées à partir d'ontologies OWL, 22èmes Journées francophones d'Ingénierie des Connaissances (IC 2011), Chambéry, France

4. Matthew Nelson, L., Peterson, J., Robert Rariden, L., Sen, R.: Transitioning to a business rule management service model: case studies from the property and casualty insurance industry. Inf. Manage. (2009). www.elsevier.com/locate/im, doi:10.1016/j.im.2009.09.007

5. Guissé, A., Lévy, F., Nazarenko, A.: From regulatory texts to BRMS: how to guide the acquisition of business rules? In: Bikakis, A., Giurca, A. (eds.) RuleML 2012. LNCS, vol. 7438, pp. 77–91. Springer, Heidelberg (2012)

6. Ram, S., Liu, J.: An agent-based approach for sourcing business rules in supply chain management. Int. J. Intell. Inf. Technol. 1(1), 1–6 (2005)

7. Guissé, A., Lévy, F., Omrane, N., Nazarenko, A., Szulman, S.: Une architecture pour la construction de modèles de règles métiers documentés. In: IC 2012 (2012)

8. Omrane, N., Nazarenko, A., Rosina, P., Szulman, S., Westphal, C.: Lexicalized ontology for the management of business rules an industrial experiment. Published in Workshop "Ontology and Lexicon" of the 9th International Conference on Terminology and Artificial Intelligence, Paris, France (2011)

9. Chniti, A., Albert, P., Charlet, J.: A loose coupling approach for combining owl ontologies and business rules. In: RuleML2012@ECAI Challenge, at the 6th International Symposium on Rules Research Based and Industry Focused 2012. CEUR Workshop Proceedings, vol. 874, pp. 103–110 (2012a)

10. Duda, J., Stawowy, A.: A possibility of business rules application in production planning. Arch. Foundry Eng. 10(2), 27–32 (2010). ISSN: 1897-3310

11. Sad Houari, N., Taghezout, N.: Towards a new agent based approach for modeling the business rules processes in small and medium enterprise in Algeria. In: 2nd International Conference on Networking and Advanced Systems (ICNAS 2015), Annaba, Algeria, 6–7 May 2015 (2015). ISBN: 978-9931-9142-0-4

12. Sad Houari, N., Taghezout, N.: A combined use between rules, ontology and agents in BRMS design: application to SME in Algeria. In: International Conference on Artificial Intelligence, Energy and Manufacturing Engineering (ICAEME 2015), Dubai, 7–8 January 2015 (2015). ISBN: 978-93-84422-05-9

13. Cryptography definition, consulted by 24/02/2015 at 14:15. http://searchsoftwarequality. techtarget.com/definition/cryptography

14. Sad Houari, N., Taghezout, N.: An agent based approach for security integration in business rules management system. In: International Conference on Intelligent Information Processing, Security and Advanced Communication (IPAC 2015), Batna, Algeria, 23–25 November 2015 (2015). ISBN: 978-1-4503-3458-7

15. Bajec, M., Krisper, M.: A methodology and tool support for managing business rules in organisations. Inf. Syst. 30, 423–443 (2005)

16. Chniti, A., Albert, P., Charlet, J.: Gestion des dépendances et des interactions entre ontologies et règles métier. Ph.D. thesis, Pierre et Marie Curie University - Paris VI (2013)

17. Terminae method definition, consulted by 24/06/2015 at 16:03. http://ontorule-project.eu/ news/news/terminae.html

18. Su, C.J., Peng, C.W.: Multi-agent ontology based web 2.0 platform for medical rehabilitation. Department of Industrial Engineering and Management, Yuan Ze University, 135, Far East Rd, Chung Li, Taiwan, ROC (2011)

Knowledge Sharing and Innovative Corporate Strategies in Collaborative Relationships: The Potential of Open Strategy in Business Ecosystems

Anna Wulf[✉] and Lynne Butel

Plymouth University, Business School, Mast House, 24 Sutton Road, Plymouth, PL4 0HJ, UK
Anna.wulf@plymouth.ac.uk

Abstract. Knowledge is a central resource in gaining competitive advantage. Sharing of knowledge between partners in collaboration has been an important research focus in the area of strategic management. In different collaborative structures, the determinants and capabilities knowledge sharing differ, as do the strategies employed, the positions taken and the roles played. The following conceptual work provides an insight into how knowledge is shared between partners, how knowledge is influenced by the partners' environment and their capabilities; depending the position they take and the roles they play.

Keywords: Networks · Business ecosystems · Competition · Collaboration · Knowledge sharing · Open strategy

1 Introduction

The topic of collaboration between companies and firms working in a network of interconnections is popular for many years now [1–4]. Resource sharing and exchange is a major reason for collaboration between partners [5]. Knowledge as a central resource to build competitive advantage and develop innovative ideas [3, 4] is considered to be the key resource for cooperation between partners [5]. When analyzing network governance and firm governance mechanisms, as well as network characteristics, the question arises if knowledge between partners can be similarly shared as knowledge within companies [6]. Similar to mechanisms in companies, knowledge between partners is shared through formal and informal relations [7]. Formal relations can be governed in a different way from informal relations and underlie distinct knowledge sharing mechanisms [7, 8]. Companies acting in networks maintain different partnerships and different relations, depending on the industry structure as well as the position fulfilled within the network [9, 10].

In this work theories about knowledge sharing within organizations, formal and informal knowledge sharing mechanisms [11, 12], knowledge sharing in networks and governance mechanisms of networks [3, 13, 14] are conceptualized by addressing findings from a combined business ecosystem and network theory approach, in order to further develop the understanding of knowledge sharing. The resource based view and resource exchange between firms determined by the type of relationship maintained [15],

© Springer International Publishing Switzerland 2016
S. Liu et al. (Eds.): ICDSST 2016, LNBIP 250, pp. 165–181, 2016.
DOI: 10.1007/978-3-319-32877-5_13

structural and relational embeddedness of the social capital perspective [16, 17] is also employed. This conceptual approach to knowledge sharing is new as the combination of business ecosystem theory and network theory helps to define and explore knowledge sharing mechanisms and capabilities in certain network environments, which influence and are influenced by the distinct roles and positions within networks.

2   Methodology

This paper provides a conceptual approach to different theories about inter-organizational relations, detecting similarities between approaches and combining them to provide a new conceptual approach network theory [18]. The paper is uses theoretical and conjectural writing to formulate new constructs to allow advanced thinking [19]. This paper offers a new approach and a new way of conceptualizing knowledge sharing amongst organizations, which has not been addressed in previous literature [18]. It is founded in the combination of business ecosystem and network theory literature and synthesizes the existing ideas to produce a new conceptual model or framework [18, 20]. In this way a new insight into the widely acknowledged issue of knowledge transfer and competitive advantage is developed [20]. Furthermore, it can be seen as a certain research strategy that contains subjectivist and interpretivist elements often driven by an author's approach to theory [20]. Conceptual research uses data, gained mainly from existing knowledge and concepts by detecting new contexts and relations [18, 20] and is defined by a topic rather than by a certain method [20, 21]. Conceptualizing research and rebuilding existing theories through a conceptual approach is very important and complements empirical research [22]. Theoretical contribution in conceptual research means that the work must offer valuable insight into the phenomenon and advance the knowledge related to it [23] by combining knowledge rather than data [24].

This study aims to re-conceptualize common elements in research of networks by philosophical conceptualization, in order to link them to Business ecosystem theory and explain the need of identifying different roles and strategies of companies acting in different network environments and stimulate theory development in that area.

3   Knowledge Sharing in Networks

Collaborative relationships often develop in order to share resources [15, 25] and mutually develop innovative ideas in order to gain competitive advantage [3]. Consequently, firms acting in collaborations cannot be understood without understanding structural and relational mechanisms in their networks [15]. Networks are seen as being neither traditional markets nor hierarchies [26]. Hierarchies such as organizational structures have different knowledge sharing mechanisms from networks and are totally different structures [12, 14]. Organizational theory investigates the mechanisms taking place within organizations that are driven by hierarchical structures and bureaucratic processes [13, 27]. Network structures lack the coordination function that organizations have [13]. As networks are neither market nor hierarchies [27] they have different governance mechanisms that contrast with normal market and pure hierarchy mechanisms [28].

"Network governance overcomes these problems (of bureaucracy) by using social mechanisms rather than authority, bureaucratic rules, standardization, or legal recourse [28]." Consequently, collaborative relationships cannot only be seen as networks but also as constructs that are determined by structural and relational factors influenced by organizational theory [12, 27, 29, 30] as well as social and business network theory [3, 14, 15, 31, 32].

Furthermore, when investigating resource exchange between different partners, the resource based view plays an essential role to explain the distinct incentives to collaborate in networks [15]. This view helps to explain resource dependency, core competencies and the importance of reaching competitive advantage [25]. Knowledge is considered to be the most important resource as it enables firms to develop new capabilities and innovative strategies [3, 13, 32, 33] and therefore plays a vital role in collaborative relationships [34]. Knowledge creation, assimilation and transfer is seen as a key capability which in turn is essential for building competitive advantage [35]. Knowledge sharing within networks cannot be done by the same mechanisms as within hierarchical structures. Hierarchy as a bureaucratic system typically relies on rules, routines and directive for authority execution but hierarchy can also be used for information and knowledge passing [33]. Networks lack these mechanisms completely or partly, depending on their network structure, and have to be governed by social network mechanisms [15] when formal mechanisms are not at present. This means that networks can be structured by formal [14] and informal relations [15]. Depending on the type of relationships maintained by the single firm of the network, different resource exchange mechanisms take place [3, 32, 36]. Grant recognized three alternatives for knowledge transfer and integration in order to share knowledge related resources which he explained to be "internalization within the firm, market contracts, and relational contacts [13]." Resource exchange in formal relations is based on saving transaction costs and enabling activities within the value chain of the company [14] which is a strong formal structure determined and controlled by formal and informal relations [13, 14, 32]. Resource exchange in social network refers to the social embeddedness perspective [37] and social capital perspective [38, 39]. Network structures are in fact the way relationships are build up [14]. Often knowledge sharing and transfer activities across intraorganisational boundaries are shaped by both, formal and informal relations [7] as it is determined by the type of relationships the company maintains [3, 32, 36].

3.1 Formal Structures, Informal Structures and Knowledge Sharing

Different researchers and different research streams concentrate on different aspects of knowledge sharing in hierarchies and networks. Summarizing the above, knowledge sharing mechanisms can be influenced by strong hierarchical formal structures as well as informal structures determined by social mechanisms.

Formal relations in networks are related to bureaucratic structures as they are common in organizational structures. Not all networks have the same governance mechanisms, therefore the ability of a single firm to access and share knowledge within a network, cannot be explained completely by formal governance mechanisms such as in hierarchies [27, 28]. Nevertheless, networks can, depending on their structural characteristics, be formed by formal contractual relationships [40]. As explained above, knowledge sharing in bureaucratic structures is different when compared to knowledge sharing in socially determined environments. The less hierarchically a network is structured, the less bureaucracy it contains and the more social mechanisms are governing the network [28]. Companies acting in a very structured and controlled network have strong governance mechanisms and they are often adapted to relatively stable conditions being a more mechanistic type of organization. The organization used to a changing environment, and less able to control its network partners, is a more organic type of organization [41]. These company and network structures will again influence the company culture, its ability to collaborate and the ability to share knowledge [12]. Hierarchy can be useful for information and knowledge passing [33] as information and knowledge sharing in hierarchies is not based on mutual information exchange due to existent authority relations [33]. In order to incorporate the knowledge gained the organization need to learn through new development and routine [33] and tacit knowledge can often be found in such routines as they are repetitive and highly patterned [42].

As described above informal relations, their type and their structure play a vital role in networks. Jarillo for example used the structural and the relational perspective to explain different structures in strategic networks [14]. The same structural approach can be taken when looking at different network structures from a social capital perspective coming from a social network background [15]. This perspective refers to a structural and relational embeddedness perspective of social and informal structures explaining resource exchange and co-creation between firms and how structural and social links between firms can enhance resource exchange determined by trust as well as a shared vision [15, 43, 44]. In order to analyze the relational setting between firms again a structural and relational dimension needs to be taken [15]. Inkpen and Tsang introduced the structured- unstructured dimension to explain different types of relationships in a network [45]. Shafique related the type of embeddedness of the firm directly to its ability to access knowledge and reach diversity and establish linkages to the knowledge base of other firms believing a greater variety of knowledge offers more possibilities to gain new knowledge [46].

Structural embeddedness refers to the degree of centrality of the company within the social network and the informal structures it acts in [47–49]. This in turn is influenced by and influences the number of informal relations and the type of informal relations maintained. Strong ties are determined by strong mutual and frequent interactions [15] which are maintained over time. The structural embeddedness of the firm [36, 47, 48] also corresponds to the view of open and sparse networks [16] and closed networks [17] defining the structure of different relationships as close or open relations [16, 17]. Not only the structural dimension is important, especially when looking at informal relations as they are not governed by hierarchical mechanisms, but by trust, mutuality and frequency of interaction [12, 30, 36, 48]. When looking at relational embeddedness, the degree of interaction and the amount of trust becomes essential which is identified by researchers to differing degrees in strong and weak ties [34, 50, 51]. The social capital perspective corresponds very strongly to Burt's, Granovetter's and Colemann's concept of strong and weak ties as well as to their closed and sparse network perspective [16, 17, 37]. Strong ties are characterized by a strong interaction, with the facilitated sharing of information and tacit knowledge [34] as the partnership is based on a mutual and deeper understanding. Network members are dependent on each other and develop trust [50]. Strong ties are also characterized by strong social control methods [51]. Weak ties, described by Granovetter as 'local bridges' [37] are not that strongly connected but more likely to deliver new information and knowledge. Weak ties therefore relate to Burt's idea of structural holes [52]. A sparse structure, observed between weak ties allows access to new and previously undetected knowledge [16]. The same refers to McEvily's and Zaheer's concept of bridging ties, as they refer to structural and relational aspects as well [53].

Some researchers tend to explain business relationships with social mechanisms. Social relationships and social mechanisms are important to develop ties and governmental structures in informal relations and they are essential for the facilitation of knowledge sharing. However, interpersonal ties differ from interorganisational ties [48, 52] as they differ from formal structures that are governed by contracts and stronger hierarchical mechanisms [12].

3.2 Network Governance and Knowledge Sharing

Summarizing the above, structural determinants in terms of formal and informal relationships are maintained as well as relational determinants referring to the degree of embeddedness of social networks can be identified as being important influencing factors to the ability to govern the network and access knowledge.

The number of formal and informal relationships in networks affects the degree of influence among network partners as formal relations can be regulated by contractual and hierarchical relations [48]. In comparison to a bureaucracy the influence and regulation is less, still it is determined by formal requirements [14]. In the same way the influence changes between partners, depending on formal and informal relations and the ability to access knowledge differs as it cannot be accessed in the same way. Explicit knowledge is easy to access and use, whereas tacit knowledge is nearly impossible to access, but only usable and learnable within its context [54, 55]. For example by learning its routines and by application [33] which makes it transfer slow, costly and uncertain [54] and which also requires strong and reliable relationships which can be either passed on a contract and hierarchy or strong informal relationship [56]. Especially tacit knowledge is rooted in procedures, norms and rules which can only be shared over time by learning from network members [57]. Therefore, specialized and tacit knowledge can be found in learning mechanisms and routines [33, 42] non-specialized and less codified knowledge in less authority based relations [3, 32, 33].

Research also found that, in clusters, proximity facilitates the exchange of tacit based knowledge and experience and can replace formal relations [58]. This corresponds to the embeddedness perspective focusing on the type of informal relations, where direct and strong ties are seen as being closely linked to the organization whereas indirect ties and weaker ties are more remote and not that close to the network [36, 47, 48, 59]. Here again the amount of influence differs by type of relationship and therefore has an impact on the knowledge shared between partners. Direct ties or dyadic ties [59] are highly influenced by solidarity, cooperation [60, 61] and trust [36]. Trust can be seen as a key factor influencing relational embeddedness fostering collaboration and knowledge sharing [48]. Another important aspect is the degree of commitment of the partners, the overlapping of objectives and maintenance effort put into the relationship [48]. Granovetter as well argues that intensity and intimacy within a network can have strong effects on resources exchanged [37]. In his concept about relational embeddedness he differs between strong and weak ties, first being strong relationship characterized with trust and detailed information exchange [34, 50, 51]. Strong and weak ties as well as direct and indirect ties therefore again describe the degree of influence partners have on each other and in turn affect the type of knowledge shared.

When looking at Burt's and Coleman's concept of structural embeddedness the degree of influence between partners becomes again important. Coleman argues that strongly embedded and closed network structures are superior to more open networks [17, 62]. Densely embedded networks with many connections and well developed social structures are seen as 'closed networks' or 'closed communities' with stronger rules of interaction [17]. Having a better control of the outcome of the network and a more structured communication, the social capital in such closed network is more beneficial and can be better used than in open networks [5, 17, 62].

Burt sees more benefits offered by networks that are not densely tied to each other which offers a greater variety and a more open approach to networks, being sparse networks [16]. Grant as well argues for as well a wider set of linkages to profit from organizations [13]. The diversity reached by collaboration can help companies to get more and diverse knowledge for an innovative use of knowledge [47, 63]. This approach to networks as open networks [5] is named sparse [52] or for example disconnected network structure [5].

The unstructured dimension refers primarily to the centrality of certain members within a network, making clear member roles and relationships possible in a very structured and centralized network and a more diffuse structure without clear tasks in an unstructured and decentralized network [13, 14, 45]. Therefore centrality and openness of the network, as well as the type of relationships among partners highly influences the mechanisms of network governance and knowledge sharing [5, 15, 28, 52]. Some researchers have investigated governance mechanisms and the type of knowledge exchanged. Sawhney and Nambisan for example differ between a centralized governance structure, determined by formal structures and hierarchical mechanisms and a community led structure being influenced by informal structures [64]. From their view, knowledge space can either be less defined and unstructured, therefore suitable for knowledge exploration [57] or very well defined and specialized suitable for exploitation [57, 64]. This knowledge in turn can be access differently depending on the structure and relations the company is acting in [57].

Summarizing the above, structural and relational determinants influence the degree of formal or informal governance mechanisms and therefore as well the mechanisms of network governance and knowledge sharing [5, 28, 52]. Networks determined by strong formal structures can act more like bureaucracies [28] exchanging knowledge differently from networks determined by more informal relations [3, 14, 32]. The degree of informal and formal relationships also refers to a more open or closed network [14, 40] comparable to Burt's and Coleman's approach to open and closed networks coming from the social network perspective and the degree of relational and structural embeddedness [16, 17, 43]. Therefore, the degree of social embeddedness and the openness of the network in terms has a strong influence on network governance mechanisms. Furthermore, the degree of openness, structural and relational embeddedness is in turn dependent on the environment that influences network structure [11, 40].

3.3 Network Structure and Its Environment

Gulati referred to networks as relational models that do not see organizations as atomistic firms but as participants embedded in closely connected industry structures that influence the nature of competition [61] and highly influenced by its dynamic environment. The influence of the environment can either be the industry or technology the company is in or its evolutionary stage which in turn influences if the network is stable and mature or unstable and developing [11, 40]. For example, perceived uncertainty in industries can hinder innovation especially in the first stage of innovation [65] and especially in early stages of innovation a great stock of knowledge can help to reduce uncertainty [66] which in turn requires different relationships to gain that knowledge than in stable environments [11].

Depending on the environment the company is in, being 'stable or variable [67]', 'low or high velocity [68]', within 'smooth or abrupt development [69]' the challenges are different [70] as well as the knowledge required [36]. This requires a more holistic view of companies acting in networks and how the networks are shaped by the environment as well as the knowledge sharing mechanisms maintained by certain actors within the networks [9, 71].

Therefore, the environment of the network can significantly influence the type of relationships maintained and the structure of the network within the industry. This again can be explained by the amount of influence needed among network partners within certain industry environments. Adner and Kapoor see the challenges faced by companies in networks relative to the position of the network in industry and the challenges the industry holds for the network [70]. Furthermore, network resources are distributed heterogeneously in networks [15], so the access to the resources can be determined by the type of relationship and the structure of the network [72]. Central positions for example seem to have a beneficial position for resource exchange, information and knowledge sharing [73–76]. Companies acting in networks can only access the resources that are available through their patterns of ties which implies that some positions in networks might be superior in knowledge sharing as they offer a different access to ties [52, 62, 77]. Network position can also influence firm behavior and outcomes [5].

Summarizing the above, network environment and network structures are determined by environmental dynamics influencing as well positions taken within networks. Business ecosystem theory focuses not only on environmental influencing factors but also on positions and strategies taken in different network surroundings.

4   Business Ecosystem Theory

Business ecosystem theory addresses the reaction of companies to a stable or fast changing environment [70] as well as the importance of certain network positions [9] and their network capabilities [11]. Business ecosystem theory couples the changing environment with the organization acting within the ecosystem, which means that external variety within the ecosystem environment also leads to internal diversity of network structure and the structure of the individual organization [78]. Therefore business ecosystem theory refers to the structural and relational dimension described above, seeing organizations as being embedded in a network of ties and social relations [61, 79] with different structural properties [16, 17] or tie attributes [37] depending on the environmental influences.

4.1   Business Ecosystems as Networks

"The business ecosystem perspective offers a new way to obtain a holistic view of the business network and the relationships and mechanisms that are shaping it, while including the roles and strategies of the individual actors that are a part of these networks [71]." Business ecosystems can be seen as open systems in which companies mutually interact with each other in order to exchange resources [80, 81]. The terminology and idea of business ecosystems come from biological systems [82, 83] and subscribe the interdependency of ecosystem actors performing different roles in order to keep the community healthy [80, 84]. Business ecosystem theory therefore offers a new perspective to approach interacting network partners and their behavior [85] being bound together by a mutual aim or a shared vision [9]. Overall, the biological metaphor was introduced to describe the idea of firms acting within and being dependent on its environment in order to meet today's challenging demands to the single firm [85]. The view of a single isolated firm acting in a market or industry between and against its competitor is complemented here by a network approach, seeing firms as being mutually dependent [63], co-evolving with each other [82, 86–88].

When investigating what authors did in terms of ecosystem structure the first aspect is the type of relationships that are maintained in business ecosystems and in what structure they are organized. Iansiti and Levien point out that: "The robustness and long-term dynamism achieved by these networks in nature highlight their power. Moreover, the specific features of theses ecosystems- their structure, the relationships among members, the kinds of connections among them, and the differing roles played by their members-suggest important analogies for understanding business networks [9]". Both authors also advance that the balance of relationships, how they are tied to each other, how the ties look like and how exactly the members are dependent on each other should be investigated further [9].

Seeing organizations embedded in network structures helps to explain resource and knowledge sharing mechanisms, still it misses the heterogeneity aspect that inherits any organization and any network. Every organization is shaped by its capabilities, abilities and structure and is therefore better or less able to access and share knowledge on the basis of its capabilities [87]. Furthermore, it inhabits a certain position of structural and relational embeddedness, with a number of ties in a dense or open network. All these differences point out that not every organization profits the same way of being embedded in network structures and that the ability to share knowledge must depend of many individual factors. Research on embeddedness has been focused on networks acting as a whole and not on what single actors can achieve or contribute within the network [48]. The focus has been to explain behavior and outcomes of networks rather than certain positions [48, 77, 89, 90].

Network resources are distributed heterogeneously within the networks and enable different access to different positions [77]. The social capital perspective already refers to the difference between actors in terms of their structural and relational embeddedness, still difference of single actors within the network remain unclear [48, 91].

4.2   Knowledge Sharing in Business Ecosystems

Business ecosystem theory investigates different roles and positions taken in a network of collaborations as well as how business ecosystem structure can influence the positions and roles fulfilled [9]. Business ecosystem structure and the position of certain players within it again influences the relationships maintained between the partners and vice versa [11, 40]. The ability to share knowledge between partners in turn, is influenced significantly by the partnerships maintained and available within the business ecosystem and the way the connection is characterized [7]. Knowledge sharing as a resource exchange and the capability to share knowledge on the level of the single company is therefore highly dependent on the environment it is in, the position and role it takes within a network, the relationships it maintains and its own characteristics [7, 9–11, 40]. Depending on knowledge sharing capabilities, companies develop the potential of their open approach and consequently their open strategy differs. The degree of openness refers here to the structural and relational openness determining the degree and type of knowledge shared in order to gain competitive advantage and generate innovative business strategies [3, 4, 92]. Therefore, as already explained above, business ecosystem theory shall help to outline how company environment, position and roles taken and relationships maintained can influence the way knowledge is shared and what kind of knowledge is shared between partners by referring to network theory and clusters, business ecosystem theory and knowledge sharing mechanisms approached from a resource based view perspective.

As described above, resulting from differences in structure and relations, not all actors within a network can fulfil the same role or occupy the same position within a network of players. This brings up the idea of different roles in business ecosystems, which has been introduced by Iansiti and Levien as well. "Perhaps the most important one is that the structure of biological networks is not homogenous [...] [9]." The roles identified by both authors refer to certain strategic roles that some authors define as ecosystem strategies [93]. Iansiti and Levien identified Keystones, Dominators, niche players and hub landlords as possible roles, which might vary or change over time [9]. The roles are explained as follows: "[...] active Keystones whose interests are aligned with those of the ecosystem as a whole and who serve as critical regulators of ecosystem health [...] embodied in a special member of the system or encoded in universally agreed-to protocols, rules, and goals- that enhances stability, predictability, and other measures of system health by regulating connections and creating stable and predictable platforms on which other network members can rely [9]." Niche players in turn are often located at the edge of the ecosystem, to bring in new ideas and innovations, whereas dominator and landlords specialize to extract value and resources out of their business ecosystems [9].

In network theory centralization is defined by the concentration of links around a focal point [29] that is why hubs are also often called focal companies [31]. Therefore a hub or focal point is located in a central position within the network as it has more connections to network members than the other firms in the network so its positions is relative to the other members [73]. The focal firm can then decide to play a dominator role or a Keystone or even a niche player role, when it is building up its ecosystem [11]. For ecosystem structures roles seem to play a vital role for the way how interconnections are build up, relationships are maintained and structured. The roles seem not be fixed to certain network positions but seem to change over time with the evolution of the business ecosystem [11, 40]. Critics of the roles explained by Iansiti and Levien are that they focus on firm level strategy rather than business ecosystem level and that the transformation and evolution of roles has not been investigated [85].

Not only the roles played are important but also the relationship between the roles play are vital to be recognized as the interactions for co-evolvement may be competitive, co-operative or co-opetitive [78]. Co-opetition means competitive cooperation [94]. The relationship between the roles is therefore as important as the role itself as it influences the development of the system itself [95] and it enables the single firm to think about its own strategic movements [10].

4.3   Ecosystem Roles and Knowledge Sharing Capabilities

As Keystones fulfil a value distribution strategy and Niche Players a value creation strategy [95] within their business ecosystem, these two most important roles shall be explained below. The physical size of a Keystone is relatively small in comparison to the population of firms within that business ecosystem [9, 88]. In order to maintain their connections and be able to distribute value, Keystone often introduce a platform of interaction for all partners of the business ecosystem [95]. The distribution of value is not an altruistic strategy but is done for the purpose of growing the own business together with the business ecosystems being the firms most important environment [9, 82, 88, 96, 97].

The platform architecture highly influences the architecture of the ecosystem, this is why a Keystone organization must consider future changes and challenges when enhancing interaction and value sharing [88, 98]. This means that poor control mechanisms and poor exchange mechanisms that do not enhance the exchange between niche players bringing in ideas for Keystones or that do not support the ability of the Keystone to communicate its needs can harm the health of a business ecosystem [9, 88, 97]. This is as the exchange of value is vital to attract new players to the ecosystem and enhance its development [82]. Niche Players usually opt for specialization that encourages them to innovate in order to maintain a sufficient level of differentiation compared to the other actors, therefore ensuring their survival. However, if platforms represent opportunities that offer access to certain resources to which they can add value through new services, they can also be construed as threats to their own survival. Indeed, if they are too generic, their services may be incorporated into the platform by the Keystone as a way for it to enhance its own value proposition [95]. Additionally, Niche Players compete within their own sub-industry in order to be able to offer the better product [88]. "[...] it is precisely this competition that keeps the ecosystem healthy: Without Niche Players who understand and exercise this leverage, ecosystems will be less healthy than they could be and may fall into sickness if their Keystones loses sight of its role [9]." Niche Players are more effective when they are independent venture as they are better in integrating new ideas and innovations as they look for them across boundaries rather than within the own boundaries [10].

Keystones and Niche Players not being aware of the structure of their business ecosystem, the network governance mechanisms in place and the organization structure necessary to face the requirements can do great harm to the health of the business ecosystem [9, 88, 97].

5   Discussion

This conceptual work combines findings from network and business ecosystem theory in order to understand the importance of the sole actor within a certain network environment influencing the whole network ecosystem. The description of Keystones and Niche Players in Business ecosystem theory relates to certain roles played in an ecosystem depending on certain positions occupied [95]. Research on network theory and network structures has concentrated on the degree of embeddedness of companies [6] rather than on the roles and positions played within networks [9] and how these roles are influenced by formal and informal relations as well as centralized and decentralized networks.

When combining network and business ecosystem theory it becomes obvious that knowledge sharing capabilities are very much influenced by the structural and relational embeddedness of the Keystone and the Niche Player [16, 17, 47–49] as well as their organizational fit to the business ecosystem requirements [9]. Depending on the degree of hierarchies in the network, the location of the Keystone as a central firm in between formal relations, being less embedded and in between formal and informal relations being highly embedded the ability to share knowledge differs [64]. The same applies to the organizational requirements set by the environment of the business ecosystem as well as the organization's capabilities to share knowledge [12]. So far, network embeddedness has been researched from a network level perspective focusing on network as a whole and not on the role of the sole actor [48]. The focus has been to explain behavior and outcomes of networks rather than certain positions [48, 77, 89, 90]. Firm strategy and its capabilities are determined by its position within the network and this in turn is influenced by the relationships maintained and the network environment. Figure 1 below describes these influencing factors and their impact on network governance mechanisms.

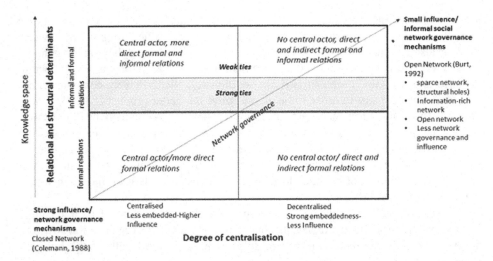

Fig. 1. Changes in network governance depending on the business ecosystem structure (own figure). The figure shows that Keystones in more closed networks need to have different knowledge sharing capabilities than Keystones in more open network structures. Their ability to access knowledge needs to be different as well, as tacit knowledge is accessed differently than explicit knowledge (see Sect. 3).

As explained above in Sect. 3 formal and informal relation determine the type of relationships maintained as well as the degree of structural and relational embeddedness. This in turn influences the degree of centralization of actors, all being influenced by the environment the network acts in. This influences again relations the actors can maintain and the position they are in and the way their organization structure is build up being more organic or mechanistic [41].

Central positions in controlled and closed network seem to better able to access knowledge as the control is higher [73–76] but this does not automatically mean that the knowledge accessible is the knowledge needed to create innovative strategies or competitive advantage. Relations in closed networks are build up differently than in open networks and enable a central firm to play a dominating role in terms of knowledge shared [99]. Less controlled and open networks can reach a higher diversity and a greater variety of linkages [46] and knowledge exploration instead of exploitation [57, 100].

Keystone and Niche Player strategy therefore need to be adjusted very well to the relationships available in their business ecosystem, the position they are in and consequently the strategy they follow. The same applies for their knowledge sharing strategy and other capabilities [53, 77]. This is, as the position again influences the possibility to develop different knowledge based capabilities [47]. Figure 2 shows the relation between knowledge available, openness of network and network governance mechanisms in place.

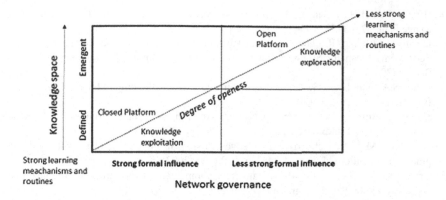

Fig. 2. Business ecosystem structures, platform openness, network governance and knowledge space (own figure)

A Keystone position might be superior in accessing knowledge as they are supposed to be deeply connected to the other actors of the network [52, 62, 77, 78] in order to drive the ecosystem [9]. Still, their capabilities and their knowledge sharing strategy needs to differ by the relationships available and maintained [7, 9–11, 40] and consequently their approach to open strategy and the knowledge accessible differs [101]. Business ecosystem theory is a broader approach to network theory that enables the company to act with its broader environment and be more open in terms of company boundaries. When there is to make strategic sense of that openness, the mechanisms of knowledge sharing in certain network environments [102] and in certain network positions as well as the capabilities required to share knowledge need to be understood. This study aims to make a first step into the direction of a concept for knowledge sharing determinants in network environments and for certain positions to meet strategic fit [103]. The next step is the investigation of knowledge sharing capabilities of Keystones and Niche Players and the influencing factors of network relations and structures within the network they act in.

References

1. McKiernan, P.: Strategy past; strategy futures. Long Range Plan. **30**, 790–798 (1997)
2. Hamel, G., Doz, Y.I., Prahalad, C.K.: Collaborate with your competitors. Harv. Bus. Rev. **67**, 133–139 (1989)
3. Lorenzoni, G., Baden-Fuller, C.: Creating a strategic center to manage a web of partners'. Calif. Manag. Rev. **37**, 146–163 (1995)
4. Chesbrough, H.W., Appleyard, M.M.: Open innovation and strategy. Calif. Manag. Rev. **50**, 57–74 (2007)
5. Ahuja, G.: Collaboration networks, structural holes and innovation a longitudinal study. Adm. Sci. Q. **45**, 425–455 (2000)
6. De Witt, B., Meyer, R.: Strategy: Process, Content, Context. Cengage, Hampshire (2010)
7. Caimo, A., Lomi, A.: Knowledge sharing in organizations: a Bayesian analysis of the role of reciprocity and formal structure. J. Manag. **41**, 655–691 (2015)

8. Hinterhuber, H.H., Levin, B.M.: Strategic networks – the organization of the future. Long Range Plan. **27**, 43–53 (1994)
9. Iansiti, M., Levien, R.: The Keystone Advantage: What the New Dynamics of Business Ecosystems Mean for Strategy, Innovation, and Sustainability. Harvard Business Press, Boston (2004)
10. Zahra, S.A., Nambisan, S.: Entrepreneurship and strategic thinking in business ecosystems. Bus. Horiz. **55**, 219–229 (2012)
11. Shang, T.: Business ecosystem capabilities: explorations of the emerging electric vehicle industry. Thesis, Cambridge University, Cambridge (2014)
12. Goh, S.C.: Managing effective knowledge transfer: an integrative framework and some practice implications. J. Knowl. Manag. **6**, 23–30 (2002)
13. Grant, R.M.: Prospering in dynamically-competitive environments: organizational capability a knowledge integration. Organ. Sci. **7**, 375–387 (1996)
14. Jarillo, J.C.: On strategic networks. Strateg. Manag. Jour. **9**, 31–41 (1988)
15. Pulles, N.J., Schiele, H.: Social capital determinants of preferential resource allocation in regional clusters. Manag. Rev. **24**, 96–113 (2013)
16. Burt, R.S.: Structural Holes. Harvard University Press, Cambridge (1992)
17. Coleman, J.S.: Social capital in the creation of human capital. Am. J. Sociol. **94**, 95 (1988)
18. Watts, R.E.: Developing a conceptual article for publication in counseling journals. J. Couns. Dev. **89**, 308–312 (2011)
19. Salomone, P.R.: Trade secrets for crafting a conceptual article. J. Couns. Dev. **72**, 73–76 (1993)
20. Xin, S., Tribe, J., Chambers, D.: Conceptual research in tourism. Ann. Tourism Res. **41**, 66–88 (2012)
21. Leuzinger-Bohleber, M., Fischmann, T.: What is conceptual research in psychoanalysis? Int. J. Psychoanal. **87**, 1355–1386 (2006)
22. Meredith, J.: Theory building through conceptual methods. Int. J. Oper. Prod. Manag. **13**, 3–11 (1993)
23. Rindova, V.: Moving from ideas to a theoretical contribution: comments on the process of developing theory in organizational research. J. Supply Chain Manag. **47**, 19–21 (2011)
24. Jabareen, Y.R.: Building a conceptual framework: philosophy, definitions, and procedure. Int. J. Qual. Methods **8**, 49–62 (2009)
25. Barney, J.: Firms resources and sustained competitive advantage. J. Manag. **17**, 99–120 (1991)
26. Thorelli, H.B.: Networks: between markets and hierarchies. Strateg. Manag. J. **7**, 37–51 (1986)
27. Powell, W.W.: Neither market nor hierarchy. Res. Organ. Behav. **28**, 295–336 (1990)
28. Jones, C., Hesterly, W.S., Borgatti, S.P.: A general theory of network governance: exchange conditions and social mechanisms. Acad. Manag. Rev. **22**, 911–945 (1997)
29. Freeman, L.C.: Centrality in social networks conceptual clarification. Soc. Netw. **1**, 215–239 (1978)
30. Ahuja, M.K., Carley, K.M.: Network structure in virtual organizations. Organ. Sci. **10**, 741–757 (1999)
31. Scott, J.: Social network analysis. Sage (2012)
32. Grant, R.M., Baden-Fuller, C.: A knowledge accessing theory of strategic alliances. J. Manag. Stud. **41**, 61–79 (2004)
33. Grant, R.M.: Towards a knowledge-based theory of the firm. Strateg. Manag. J. **17**, 109–122 (1996)

34. Uzzi, B.: Social structure and competition in interfirm networks: the paradox of embeddedness. Adm. Sci. Q. **35**, 35–67 (1997)
35. Quintane, E., Mitch Casselman, R., Sebastian Reiche, B., Nylund, P.A.: Innovation as a knowledge-based outcome. J. Knowl. Manag. **15**, 928–947 (2011)
36. McEvily, B., Marcus, A.: Embedded ties and the acquisition of competitive capabilities. Strateg. Manag. J. **26**, 1033–1055 (2005)
37. Granovetter, M.S.: The strength of weak ties. Am. J. Sociol. **78**, 1360–1380 (1973)
38. Adler, P.S., Kwon, S.W.: Social capital: prospects for a new concept. Acad. Manag. Rev. **27**, 17–40 (2002)
39. Portes, A.: Social capital: its origins and applications in modern sociology. Ann. Rev. Sociol. **24**, 1–24 (1998)
40. Rong, K., Hou, J., Shi, Y., Lu, Q.: From Value Chain, Supply Network, Towards Business Ecosystem (BE): Evaluating the BE Concept's. IEEE Press, Macao (2010)
41. Roffe, I.: Innovation and creativity in organisations: a review of the implications for training and development. J. Eur. Ind. Train. **23**, 224–241 (1999)
42. Winter, S.G.: Understanding dynamic capabilities. Strateg. Manag. J. **24**, 991–995 (2003)
43. Nahapiet, J., Ghoshal, S.: Social capital, intellectual capital, and the organizational advantage. Acad. Manag. Rev. **23**, 242–266 (1998)
44. Tsai, W., Ghoshal, S.: Social capital and value creation: the role of intrafirm networks. Acad. Manag. J. **41**, 464–476 (1998)
45. Inkpen, A.C., Tsang, E.W.: Social capital, networks and knowledge transfer. Acad. Manag. Rev. **30**, 146–165 (2005)
46. Shafique, M.: Thinking inside the box? Intellectual structure of the knowledge base of innovation research. Strateg. Manag. J. **34**, 62–93 (2013)
47. Zheng, S., Zhang, W., Du, J.: Knowledge-based dynamic capabilities and innovation in networked environments. J. Knowl. Manag. **15**, 1035–1051 (2011)
48. Gulati, R., Lavie, D., Madhavan, R.: How do networks matter? The performance effects of interorganizational networks. Res. Organ. Behev. **31**, 207–224 (2011)
49. Nohria, N., Eccles, R.: Problems of explanation in economic sociology. Networks and Organizations: Structure, Form, and Action. Harvard Business School, Boston (1992)
50. Larson, A.: Network dyads in entrepreneurial settings: a study of the governance of exchange relationships. Adm. Sci. Q. **37**, 76–104 (1992)
51. Krackhardt, D., Hanson, J.R.: Informal networks: the company behind the chart. Harv. Bus. Rev. **71**, 104 (1993)
52. Rowley, T., Behrens, D., Krackhardt, D.: Redundant governance structures: an analysis of structural and relational embeddedness in the steel and semiconductor industries. Strateg. Manag. J. **21**, 369–386 (2000)
53. McEvily, B., Zaheer, A.: Bridging ties: a source of firms heterogeneity in competitive capabilities. Strateg. Manag. J. **20**, 1133–1156 (1999)
54. Kogut, B., Zander, U.: Knowledge of the firm, combinative capabilities, and the replication of technology. Organ. Sci. **3**, 383–397 (1992)
55. Marabelli, M., Newell, S.: Knowledge risks in organizational networks: the practice perspective. J. Strateg. Inf. Syst. **21**, 18–30 (2012)
56. Everett, M.G., Krackhardt, D.: A second look at Krackhardt's graph theoretical dimensions of informal organizations. Soc. Netw. **34**, 159–163 (2012)
57. March, J.G.: Exploration and exploitation in organizational learning. Organ. Sci. **2**, 71–87 (1991)

58. Hoffmann, V.E., Bandeira-de-Mello, R., Molina-Morales, F.X.: Innovation and knowledge transfer in clustered interorganizational networks in Brazil. Latin Am. Bus. Rev. **12**, 143–163 (2011)
59. Granovetter, M.: Problems of Explanation in Economic Sociology. Harvard Business School Press, Boston (1992)
60. Emerson, R.M.: Power dependence relations. Am. Sociol. Rev. **27**, 31–41 (1962)
61. Gulati, R., Nickerson, J.A.: Interorganizational trust, governance choice, and exchange performance. Organ. Sci. **19**, 688–708 (2008)
62. Walker, G., Shan, W., Kogut, B.: Social capital, structural holes and the formation of an industry network. Organ. Sci. **8**, 109–125 (1997)
63. Brass, D.J., Galaskiewicz, J., Greve, H.R., Tsai, W.: Taking stock of networks and organisations: a multilevel perspective. Acad. Manag. J. **47**, 795–817 (2004)
64. Sawhney, M., Nambisan, S.: The Global Brain: Your Roadmap for Innovating Faster and Smarter in a Networked World. Pearson, Prentice Hall, Upper Saddle River (2007)
65. Meijer, I.S., Hekkert, M.P., Koppenjan, J.F.: The influence of perceived uncertainty on entrepreneurial action in emerging renewable energy technology: biomass gasification projects in the Netherlands. Energy Policy **35**, 5836–5854 (2007)
66. Matusik, S.F., Fitza, M.A.: Diversification in the venture capital industry: leveraging knowledge under uncertainty. Strateg. Manag. J. **33**, 407–426 (2012)
67. Lawrence, P.R., Lorsch, J.W., Garrison, J.S.: Organization and Environment: Managing Differentiation and Integration. Harvard University, Boston (1967)
68. Eisenhardt, K.M.: Building theories from case study research. Acad. Manag. Rev. **14**, 532–550 (1989)
69. Suarez, F.F., Lanzolla, G.: The role of environmental dynamics in building a first mover advantage theory. Acad. Manag. Rev. **32**, 377–392 (2007)
70. Adner, R., Kapoor, R.: Value creation in innovation ecosystems: how the structure of technological interdependence affects firm performance in new technology generations. Strateg. Manag. J. **31**, 306–333 (2010)
71. Anggraeni, E., Den Hartigh, E., Zegveld, M.: Business ecosystem as a perspective for studying the relations between firms and their business networks. In: ECCON 2007 Annual Meeting, Bergen aan Zee (2007)
72. Gulati, R., Gargiulo, M.: Where do interorganizational networks come from? Am. J. Sociol. **104**, 177–231 (1999)
73. Arya, B., Zhiang, L.: Understanding collaboration outcomes from an extended resource-based view perspective: the roles of organizational characteristics, partner attributes, and network structures. J. Manag. **33**, 697–723 (2007)
74. Galaskiewicz, J.: Interorganizational relations. Ann. Rev. Sociol. **11**, 281–304 (1985)
75. Ibarra, H.: Network centrality, power, and innovation involvement: determinants of technical and administrative roles. Acad. Manag. J. **36**, 471–501 (1993)
76. Powell, W.W., Kogut, B., Smith-Doerr, L.: Interorganizational collaboration and the locus of innovation: networks of learning in biotechnology. Adm. Sci. Q. **41**, 116–145 (1996)
77. Gulati, R.: Network location and learning: the influence of network resources and firm capabilities on alliance formation. Strateg. Manag. J. **26**, 397–420 (1999)
78. Peltoniemi, M., Vuori, E., Laihonen, H.: Business Ecosystem as a Tool for the Conceptualisation of the External Diversity of an Organization. Tampere University of Technology, Tampere (2005)
79. Gulati, R., Singh, H.: The architecture of cooperation: managing coordination costs and appropriation concerns in strategic alliances. Adm. Sci. Q. **43**, 781–814 (1998)

80. Garnsey, E., Leong, Y.Y.: Combining resource-based and evolutionary theory to explain the genesis of bio-networks. Ind. Innov. **15**, 669–686 (2008)
81. Scott, W.R.: Organizations. Prentice Hall, Englewood Cliffs (1987)
82. Moore, J.F.: Predators and prey: a new ecology of competition. Harv. Bus. Rev. **71**, 79–86 (1993)
83. Moore, J.: The Death of Competition: Leadership and Strategy in the Age of Business Ecosystems. Harper Business, New York (1996)
84. Li, F.F., Garnsey, E.: Building Joint Value: Ecosystem Support for Global Health Innovations, pp. 69–96. Emerald Group Publishing, Bingley (2013)
85. Rong, K., Yongjiang, S.: Business Ecosystems: Constructs, Configurations, and the Nurturing Process. Palgrave Macmillan, Basingstoke (2014)
86. Basole, R.C.: Structural analysis and visualization of ecosystems: a study of mobile device platforms. In: Proceedings of the AMCIS 2009, San Francisco (2009)
87. Teece, D.J.: Explicating dynamic capabilities: the nature and microfoundations of (sustainable) enterprise performance. Strateg. Manag. J. **28**, 1319–1350 (2007)
88. Mäkinen, S.J,, Dedehayir, O.: Business ecosystem evolution and strategic considerations: a literature review. In: 18th International Conference on CITER, Tampere (2012)
89. Gulati, R.: Alliances and networks. Strateg. Manag. J. **19**, 319–321 (1998)
90. Lavie, D.: The competitive advantage of interconnected firms: an extension of the resource based view. Acad. Manag. Rev. **31**, 638–658 (2006)
91. Aral, S., Walker, D.: Tie strength, embeddedness, and social influence: a large-scale networked experiment. Manag. Sci. **60**, 1352–1370 (2014)
92. Whittington, R., Cailluet, L., Yakis-Douglas, B.: Opening strategy: evolution of a precarious profession. Br. J. Manag. **22**, 531–544 (2011)
93. Iyer, B., Lee, C.H., Venkatraman, N.: Managing in a "small world ecosystem": lessons from the software sector. Calif. Manag. Rev. **48**, 28–47 (2006)
94. Brandenburger, A.M., Nalebuff, B.J.: Co-Opetition. Crown Business, New York (2011)
95. Isckia, T.: Amazon's evolving ecosystem: a cyber-bookstore and Application Service Provider. Can. J. Adm. Sci. **26**, 332–343 (2009)
96. Gawer, A., Cusumano, M.A.: Industry platforms and ecosystem innovation. J. Prod. Innov. Manag. **31**, 417–433 (2014)
97. Fox, P.B.: Creation and Control in Business Ecosystems. Llull (2013)
98. Tiwana, A., Konsynski, B., Bush, A.: Platform evolution: coevolution of platform architecture, governance, and environmental dynamics. Inf. Syst. Res. **21**, 675–687 (2010)
99. Epicoco, M.: Knowledge patterns and sources of leadership: mapping the semiconductor miniaturization trajectory. Res. Policy **42**, 180–195 (2013)
100. Crespo, J., Raphael, S., Jerome, V.: Lock-in or lock-out? How structural properties of knowledge networks affect regional resilience. J. Econ. Geogr. **14**(1), 199–219 (2014)
101. Connell, J., Kriz, A., Thorpe, M.: Industry clusters: an antidote for knowledge sharing and collaborative innovation? J. Knowl. Manag. **18**, 137–151 (2014)
102. Bathelt, H., Cohendet, P.: The creation of knowledge: local building, global accessing and economic development—toward an agenda. J. Econ. Geogr. **14**, 869–882 (2014)
103. Kim, T.H., Lee, J.N., Chun, J.U., Benbasat, I.: Understanding the effect of knowledge management strategies on knowledge management performance: a contingency perspective. Inf. Manag. **51**, 398–416 (2014)

DSS Technology Improving System Usability and Feasibility

Technology Improvement System
Capability of a Facility

Developing Innovative Tool to Enhance the Effectiveness of Decision Support System

Fahad Almansour$^{(\boxtimes)}$ and Liz Stuart

School of Computing and Mathematics, Plymouth University, Plymouth, UK
{fahad.almansour,liz.stuart}@plymouth.ac.uk

Abstract. This research centres on Usability Evaluation Methods (UEMSs) with the aim of supporting developers' decisions in the use of learning resources in achieving efficient usable system design. The suggestion is made pertaining to a new usability evaluation model dEv (stand for Design Evaluation) with the objective to support decisions to overcome three key obstacles: firstly, the involvement of users in the preliminary stages of the development process; (2) developers' mind set-related issues as a result of either their lack of UEMS or the provision of too many; and (3) the complete lack of understanding surrounding UEMS importance. An experimental approach was applied in addition to a survey-based questionnaire in an effort to examining the issues pertaining to UEMS. Empirical works were carried out with system developers in order to test the dEv, the results of which have been presented from the empirical study to support various considerations, such as: system developers' decisions and their involvement in the earlier phases of the design of systems; the gathering of specifications and end-users' feedback; and enhancing usability evaluation learning capacity.

Keywords: Usability · UEMS learning resource · Evaluation methods · Usable system design · Decision making

1 Introduction

Software usability is the main goal of producing products. The International Organization for Standardization (ISO 9241-11) referred to usability as '*the extent to which a product can be used by specified users to achieve specified goals with effectiveness, efficiency and satisfaction in a specific context of use*' [1]. Incredibly, usability has been incorrectly classified as a part of software development attached or added on towards the end of the development cycle; this is often misunderstood as being part of the 'finishing' of the product, which is totally incorrect. Usability is core to the success of the software. Usability is central to efficiency and throughput, and thus it is core to business and software development [2].

Measuring the usability of software is another main concept that should developers and originations concern about it and its considered as a central activity in the usability process [3]. Evaluation is the means by which evaluators assess: (i) the quality of the software, (ii) the usability of the system, (iii) the extent to which the user's requirements have been met, and (iv) the ability to identify system problems. Note that the

© Springer International Publishing Switzerland 2016
S. Liu et al. (Eds.): ICDSST 2016, LNBIP 250, pp. 185–201, 2016.
DOI: 10.1007/978-3-319-32877-5_14

latter is based on user satisfaction with the system [4, 5]. This stage of the software development lifecycle is crucial for many reasons. Firstly, the communication of requirements between users and developers is a difficult task, despite the numerous methodologies that exist to support this process. Thus, it is essential that checks are carried out so as to ensure the product matches user expectations. It is equally important to ensure that the developer has an in-depth appreciation of how the user intends to use the software, as well as any limitations of the software. Thus, it is a very worthwhile and important stage of the software development lifecycle [6].

Software developers usability and evaluation knowledge's is essential factor that lead them to consider there two concepts during the development process. Thus, learning resources a way to enhance the developer's skills and about how to measure their product and enhance their ability of making decision about the usability level [7]. Furthermore, the aim of creating educational software is to support learning. However creating an appropriate resource for novice user and support their request is challenge [8].

This paper proposes that developer knowledge of both usability and evaluation methods impact the decision making toward to friendly product for the end user that easy to use. Thus, our suggestion is aimed at bringing software usability, user involvement and software evaluation concepts together in an effort to produce usable interaction design. This combination is presented as a tool known as dEv, which represents Design Evaluation. Hence, our dEv model has built to prompt the developers for using evaluation methods during the development process in order to increase the software usability [9]. This study we measure the suggested model to investigate two points: (1) the learning resource usability issues. (2) The participates ability for making decision on the dEv learning resource. In this paper we report our findings, the emerging themes of our qualitative research conducted using user testing and thinking aloud, also using questionnaire for quantitative research. The data was collected by user testing, thinking aloud, observation and questionnaire with the developers who have already involved in the learning resource requirements.

The paper is organised as follows. The following section reviews relevant literature, while Sect. 3 presents the research methodology to evaluate the proposed model. Section 4 describes the finding of the study, followed by Sect. 5 presents a discussion of the results. Section 6 concludes the study and suggests future work.

2 Literature Review

This section takes the key software usability and usability evaluation method concepts and reviews these in mind of their ability to support the creation of usable software. Predominantly, this review centres on various usability obstacles and how these may be overcome and managed throughout the process of development. Furthermore, these obstacles impact developers' ability to make decisions of any software usability.

Design Principles Developers with different level of programming experience are able to create such as product; however creating a usable product is the challenge. Thus, number of design principles and guidelines are established to help developers creating usable software. For example Nielsen coined his own key aspects of usability when he listed the following attributes to identify the usability of a system [4, 10] present their taxonomy and various principles, such as simplicity, structure, consistency and toler-ance. A number of design principles referred to as 'Eight Golden Rules', where these 8 principles are considered as guideline for software design [11].

Following common design principles to create a usable product is still encountering number of challenges that prevent developers to ease of use product. User involvement is one challenge that impacts the software usability. Ardito et al. [13] reported that user involvement on the early stage of the development is positively impact the design, however it's complicated [12]. Furthermore, problem of not involving the user within the development of software has also been identified commercially [13]. Conducting evaluation methods during the development process also consider as challenge.

User involvement. User-centred design (UCD) is a way of improving overall software usability [14, 15]. User-centred design is a well-established software development methodology that incorporates design evaluation within the core development lifecycle [14]. There are different terms used in the field that have the same concept of UCD: for instance User-Centred System Design (UCSD), Usability Engineering [16] and Human-Centred Design (HCD) [17]. The term UCSD originates from research by Norman [18] back in the 1980s. Since then, UCSD has been widely adopted [19, 20]. It focuses on user's requirements and needs during the software development stages [6]. The concept of UCD emphasises the involvement of the user and their activities in order to achieve product goals. Thus, collecting data and information at the early stage of the development process is considered as a key element for supporting decision maker [21]. Bevan [22] lists a number of benefits behind the UCD main concept and these benefits impacted on the product sales, development cost, easy- of - use and supporting and maintaining costs [22].

The agile development process has been widely integrated with other development processes: for instance, User-Centred Design (UCD). The integration aims at improving the level of software usability by combine the strength points of both approaches and puts them in one model to solve the development challenges. For instance, user involvement is one of the challenges facing developers during the development process; thus, integration is a way of addressing this challenge. Many authors have proposed different integration frameworks for different levels of integration, where each frame-work has its own goals [23–28].

Developer Ability for Conducting Evaluation. Developer preferences may be viewed as one of the predominant obstacles hindering the development of usable software, as in the case of the mind-set of the developer. It has been stated by [13] that there are three key obstacles effecting developers' ability to create usable software and carrying out usability assessments, including development mind-set, the wealth of resources necessary to complete a usability evaluation, and the problems and com-plexities involving users in the usability evaluation process. Accordingly, various efforts have been made by developers to avoid users' participation in the development stages owing to the view that such involvement can waste time, may mean unrealistic requests, and the uncertainty of users concerning their needs [13]. Moreover, the lack of usability evaluation knowledge is one of the key issues facing developers in the completion of the usability assessments on products [29]. Accordingly, the creation of software with a lower level of usability evaluation knowledge and without the direct participation of users in the process of development can mean developers create products based on their viewpoints and own experiences. As such, some products following the completion of development.

Developer training proposed as solution to increase the developer awareness of user involvement and evaluation conduction. Thus, Number of previous studies identified that inexperienced usability evaluators are able to conduct the usability evaluation by using tools, training or learning resource to come up with list of identifying problems [30–32]. Furthermore, in 2012 Skov and Stage conducted a study to investigate the student ability of conducting evaluation after they have training course. This study provided 234 of first-year undergraduate students with 40 h of training. As results of this experiment students "gained good competence in conducting the evaluation, defining user tasks and producing a usability report, while they were less successful in acquiring skills for identifying and describing usability problems" [33].

As mentioned above, many authors have explored three common challenges facing developers throughout the development process. Each study has focused on, and accordingly attempted to solve, one issue. Following design principles is important; however, numerous design principles already established could disrupt developers from creating usable software for two causes: firstly, it is difficult for developers to decide which principles are the best to follow; and secondly, following these principles could impact the lack of evaluation process and user involvement. Additionally, user involvement also is another important concept developers should be concerned with; thus, UCD has been established and integrated with different software development process. Most of these integrations have added value to software usability. Furthermore, developers' preferences and behaviour significantly impact developers in the creation of usable design. Thus, learning resources are considered as a way of improving developers' knowledge about the evaluation methods, designing behaviours and evaluation-conducting sessions. In summary, these challenges and benefits, as shown in the literature, have driven us to combine these into one tool so as to support benefits and improve challenges. Furthermore, our suggested tool should fit in with the early stages of the design process until the deployment stage in an effort to increase the decision support system towards usability.

3   Research Method

This work applied usability assessment approaches involving end users in evaluating the design and accordingly gathering dEv-related feedback. Accordingly, there has been the use of four common approaches at various stages of the usability evaluation process, namely observations, thinking aloud, questionnaire and user-testing [34].

Table 1 details the various testing methods applied in this work, and the various objectives underpinning their use. User testing enabled participants to carry out tasks on the real dEv product. Such a method was applied as one of the key aspects in usability methods that results in users being able to interact with the real design and deliver critical data whilst also increasing usability issues. Moreover, user testing is a suitable method to establishing the capacity of users to establish usability issues and accordingly propose solutions [35]. Thinking aloud also has been applied in order to encourage participants to communicate what they are doing. Participants' thoughts throughout the user testing stage aids in the establishing of system errors and the root causes behind issues [34]. Observation was also adopted in order to focus on the interaction of users throughout the user-testing process. This method is useful in establishing the key usability issues and accordingly creating a usable user interaction design [36]. The questionnaire method was implemented in mind of gathering data pertaining to user satisfaction with our dEv resource. Questionnaires are recognised as valuable when striving to gain insight into users' feelings throughout testing and accordingly measuring their degree of product satisfaction [37]. Moreover, the questionnaire is regarded as a suitable method for gathering quantitative data to compile statistics [34]. Nielsen states that, 'the first several usability studies you perform should be qualitative' [38]; thus, we align with this idea and accordingly include three methods in an effort to gather qualitative data opposite to one quantitative method to establish the user satisfaction of the model. Both methods have been used to strengthen the results [39]; furthermore, a mix-methods approach would be useful to expand the scope of the study findings [40].

Table 1. Research method phases for the dEv usability testing

Phase No	Phase denomination	Purpose and achievement
Phase (1)	User Testing	• To assess the (dEv) learning resource • To increase usability issues • To complete measurement of users' overall capacity to enhance the model • To discuss user finding issues
Phase (2)	Thinking aloud	• To understand users thoughts throughout the testing process
Phase (3)	Observation	• To observe the interactions of users throughout the testing process • To detail additional usability issues
Phase (4)	Survey based Questionnaire	• To establish the degree of user satisfaction with the learning resource • To collect user recommendations and solutions

The sample contained 10 participants, all of whom had some degree of participation in the designing development process of the dEv. There is vast argument pertaining to the usability evaluation sample size and how many participants should be included; thus, numerous authors have come to recognise that usability evaluation does not require a large sample size. Therefore, our sample size was aligned with the suggestion made by Nielsen in regards usability evaluation concept, and further aligns with the 10 ± 2 rule [41, 42] for various reasons. First and foremost, this is the first evaluation session for the dEv model, which provides the fundamental elements for creating the dEv model.

Moreover, there also are no significant functions that require a large assessment sample. Secondly, as the literature mentioned, a large number of participants for usability evaluation would mean the repeating of the same issues already identified in a small sample of participants, which is costly. Thus, the recruited sample size was aligned with the purpose of the study.

The main criteria for the sample selection were as follows: subjects should have programming experience; and the subjects should have a minimum of one designing product and be able to learn more about the user interaction design assessment approaches. This work has carried out in-person sessions, involving individuals sitting with the researcher on a one-to-one basis. Approximately 1 h was assigned to each participant. The subjects were seen to have some degree of dEv-use experience, albeit differing: 3 subjects described themselves as 'beginners', whilst 7 were 'intermediate' and 1 'expert'.

Procedure. Each participant took approximately one hour to complete this study tasks and collect feedback. At the beginning of the session, the researcher introduced the study aims and procedure, and then asked them to sign paperwork. This study asked participants performing two tasks to achieve the study goals. These tasks aimed to preamble participants on the learning resource. There were two test scenarios, including four test cases, which were to be completed and would take approximately 20 min. Afterwards, participants were asked to choose any unknown evaluation method form the dEv resource and free testing to explore and learn from it. During the free-testing, some issues were highlighted for discussion with the participants and enabled further collection of feedback and usability problems. Furthermore, thinking-aloud conducted during the free testing part to collect more data about the participant finding issues. At the end of the session, the questionnaire was filled in an effort to measure user satisfaction with the resource.

Data Analysis. This research provides both qualitative and quantitative data. Thus, software was used to assist the data analysis process, such as SPSS for quantitative data or Computer Assisted Qualitative Data Analysis Software (CAQDAS), such as NVivo for qualitative data. In mind of the aim of this research, both SPSS and NVivo analysis software tools were applied to analyse the data (Fig. 3). Furthermore, the questionnaire data was collected by Survey Monkey, which provided an analysis system for use alongside SPSS for advance analysis.

4 Research Findings

This section's aim is to provide an overview of the evidence garnered from the key findings of the literature, as well as from the empirical works supporting the decisions made by system developers and their overall involvement in the earlier stages of system design; improving usability assessment learning capacity and the garnering of end-user feedback and user requirements (Fig. 1).

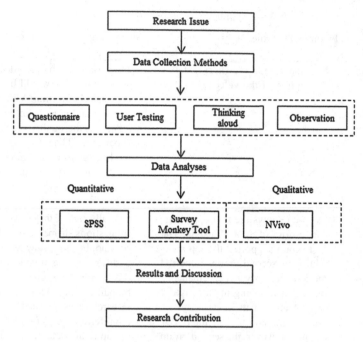

Fig. 1. Research process

4.1 System Developers' Decisions and Their Involvement in the Early Phase of the Design System

Usability Themes. An overall this study comes up with 9 usability main themes, which can be divided into 26 sub-themes. The main themes and sub-themes discussed during the users testing sessions. Table 2 shows examples of constructs with corresponding experiment session.

Table 2. Examples of constructs with corresponding experiment session

Main construct	Findings from the study	Supporting from literature
Video issues	The complete agreement of the video structure interface design means that participants were happy to have a short description about the video contain. As discussed, this description helps them to understand what this video is about, and shows that participants were happy to have a combination	The literature review has highlighted the substantial role of assistant tools used for increasing user's knowledge about related topic [43]. These different techniques clearly showed that participants have different levels of experience and viewpoints. However, all of these suggestions clearly show that long videos were not of interest

(Continued)

Table 2. (*Continued*)

Main construct	Findings from the study	Supporting from literature
	between videos and short text description of the video	and should not be included because nobody would be willing to watch them. Short videos with relevant and concise information reflect on the quality of the user experience. Though long videos might contain all the information needed, it might distract the user from the surrounding environment [44]
Menu styling issues	This study shows the two opposing suggestions of the menu style. The first group supported the top style whilst the second group supported the left menu style. Participants interested in seeing the top style was because of its modern style and its common use for interface design. However, the second group argued that the top menu would not appear all the time on the screen, meaning this will increase user actions on the interface. Furthermore, the top menu could distract the user with the browser tool bar, which may allow the user to leave the resource. Moreover, the top menu might not be clear enough for the user against the left menu style. For these reasons the majority of the participants supported the left menu for this resource. Menu style design is considered to be significant in making information on web sites easy to find	Essentially, all of these two suggestions of menu style are commonly used; however, the tradition left menu style will fit on this learning resource, as supported by participants. Although previous research has suggested a left menu is preferred by users, recent literature also claim that the selections of menu style is a personal choice as long as it is usable [45]
Navigation issues	The difference in navigation methods is important to consider on the interface design as this gives users multiple ways of controlling the interface. Therefore, links to navigations have been used on the study designing interface and mostly are accepted by	Thus, 'breadcrumb' navigation could be a solution for this [46]

(*Continued*)

Table 2. (*Continued*)

Main construct	Findings from the study	Supporting from literature
	participants. The study shows the link locations were at the right and clear place. However, extra links for more navigation were mentioned as provided: for instance, 'next' and 'previous' at the end of each page. This means participants could deal with multiple navigation methods or do this through their own design experience. The user getting lost could be the main cause of preventing the use of links for page navigation	
Texts issues	Scrolling is the default method to read or see long text on a browser. However, the study shows that some participants support short text on the interface by using the collapse and expand method '+,-'. Those participants wanted to have a clear interface design and be presented with only the important information, with an additional option to expand the information. Furthermore, this was also mentioned as unnecessary information only. All of them agreed that some information should be hidden. In this study, there were three participants who argued that all including information is important to read and should appear all the time; otherwise, some users would not read some sections or would get lost somewhere	Johnson argued various advantages and disadvantages of using scrolling on the design [47]. The combination between these two techniques is expected to be more useful and will improve usability; however, consideration is given to which information will be hidden and which information will be appeared

4.2 Enhance the Usability Evaluation Learning Capacity

The present work has identified various new themes, as highlighted by the study subjects. Table 3 shows the list of new themes identification and interpretation during the user testing sessions.

Table 3. New themes identification and interpretation

New identified themes	Interpretation of the themes identified
Lack of information	This feedback is clearly shows that some participants are interested in having extra information and additional topics, also its way increase their awareness of decision making [48]. The resource should targets both novice and expert users, thus including and advance topic should be planned and take it as a separate study. Researcher needs to review the topics and meet users in deciding how to integrate this on the main resource
Contact E-mail	This is known as utility navigation feature and is considered to be one of the activities strongly impacting user satisfaction with the design [49]. This suggestion is one of the most important feedbacks. This suggestion means the resource updating will be regularly and based of the users using. We must keep on touch with users all the time and create an email or contact form for any further suggestions and feedbacks
References design solution	This evaluation study allows users to be involved in the design process by creating suggestion to reinstruct the design. A references section is one of the sections where participants are given some examples for redesign. This means some users are willing to be involved in the design process by giving design suggestions. These suggestions are considered on the next version of the software
Text and reference integration	The integration between the contents and references is important to keep the user related with the original sources for the content. This method also will reduce the time of learning about references between the lists of references. However, this integrating could be way to distract the users with a lot of references links. The searching tool could be a solution instead of the integration, thus we should planning to add this service and well presented on the further version of the resource

4.3 Gathering of End-Users' Satisfaction and Feedback

At the end of the user testing sessions, participants were asked to complete a user satisfaction questionnaire. Table 4 shows the participants' agreed percentages in regard to the dEv interface elements. This study show that participants were in complete agreement that dEv has clear structure, easy menu style, enough content to understand the topic and the images that provided are helpful too. However, 70 % of the study participants agreed that dEv provides an easy navigation. Extra information about each topic is included as references, where these references have been placed as the part of the topic main pages. In total, 70 % of the participants rated these references as helpful references, which encouraged them to explore the topic in depth; in contrast, 10 % disagreed and 20 % were undecided. However, only 60 % agreed that references should be placed at the right position on the interface whilst 10 % of them disagreed. The study results show that 80 % of the participants agreed that the links provided were

Table 4. Percentage of agreed statements

Statements	% Agree
Clear interface structure	100 %
Easy navigation	70.0 %
Menu style	100 %
References position	60.0 %
Content structure helps me to clearly understand the presented topic	100 %
Using links through text to jump between different topics is obvious	80.0 %
References encourage me to expand the topic for more information	70.0 %
Using images is helpful to understand the topic	100 %
Using videos is helpful to understand the topic	80.0 %
The videos provided reduced learning time	60.0 %

For each of the following categories, please rate your satisfaction level with our software: | Software Interface

		Frequency	Percent	Valid Percent	Cumulative Percent
Valid	Neutral	1	10.0	10.0	10.0
	Satisfied	6	60.0	60.0	70.0
	Very Satisfied	3	30.0	30.0	100.0
	Total	10	100.0	100.0	

For each of the following categories, please rate your satisfaction level with our software: | Overall Appearance

		Frequency	Percent	Valid Percent	Cumulative Percent
Valid	Dissatisfied	1	10.0	10.0	10.0
	Neutral	1	10.0	10.0	20.0
	Satisfied	4	40.0	40.0	60.0
	Very Satisfied	4	40.0	40.0	100.0
	Total	10	100.0	100.0	

For each of the following categories, please rate your satisfaction level with our software: | Usability

		Frequency	Percent	Valid Percent	Cumulative Percent
Valid	Dissatisfied	1	10.0	10.0	10.0
	Satisfied	3	30.0	30.0	40.0
	Very Satisfied	6	60.0	60.0	100.0
	Total	10	100.0	100.0	

Fig. 2. An overall of user satisfaction

clear and easy to find; however, 20 % of them rated this as unclear. Using videos on the dEv resource were rated as a useful way of understanding the topic by 80 % of the study participants whilst the rest 20 % remained undecided. However, 60 % of the study participants agreed that the videos provided were helpful and reduced learning time, whereas 10 % were disagreed and 30 % were undecided in terms of whether or not these videos were helpful and the right choice.

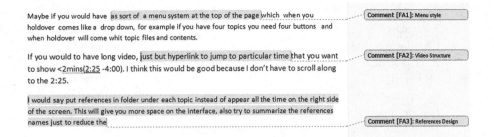

Maybe if you would have as sort of a menu system at the top of the page which when you
holdover comes like a drop down, for example if you have four topics you need four buttons and
when holdover will come whit topic files and contents.

Comment [FA1]: Menu style

If you would to have long video, just but hyperlink to jump to particular time that you want
to show <2mins(2:25 -4:00). I think this would be good because I don't have to scroll along
to the 2:25.

Comment [FA2]: Video Structure

I would say put references in folder under each topic instead of appear all the time on the right side
of the screen. This will give you more space on the interface, also try to summarize the references
names just to reduce the

Comment [FA3]: References Design

Fig. 3. Illustrates the process of identifying and noting the codes and themes

Overall satisfaction is an important goal when applying the questionnaire. The participants were asked to rate their overall satisfaction with three elements: the design of the software interface, the appearance and the usability. The results show that, overall, 90 % of the study participants were satisfied with the design of the software interface whilst 10 % were neutral. The software appearance was rated as satisfied by 80 % of the participant; however, 10 % of the participants were dissatisfied, whilst the same percentage were neutral. The majority of participants (90 %) were satisfied with the usability level of the dEv software; the rest (10 %) were dissatisfied and claim it should be improved (see Fig. 2).

5   Discussion

This study was conducted in order to evaluate the first version of the dEv resource and accordingly come up with a new improvement planning and how the awareness impacts the decision making. The dEv resource has been built based on multimedia (videos with words and images) and links (either navigation or references links). Thus, some usability problems or improvement suggestions are expected from the study participants. The following list provides more discussion on the study findings providing support for the improvement of UMES important understanding, as well as that of the user's feedback and the usability concept.

As found in this research, participants have an interest in everything being as short as possible, with most of them making suggestions or comments on the long text and videos. Participants were not completely satisfied with the length of videos. Thus, they came up with three different solutions to reduce the video time. These three suggestions clearly show that most of the study participants were unhappy with long videos, with short videos more acceptable. Learning cognitive load is valuable and should be minimised on the learning resource. [50] State that 'cognitive load is a central consideration in the design of multimedia instruction'; thus, learning cognitive load should be reduced and they summarise number of ways that can be solving for this challenge and reduce the cognitive overload. However, at times, short content is inadequate in terms of presenting the learning topic and therefore affects understanding. This short content support could affect the understanding pertaining to UEMS importance amongst developers. It is recognised that one solution provided through the application of the dEv model to support the view on content length is the inclusion of summary alongside text and video, affecting the understanding of UEMS topics amongst participants. The study results show that participants were happy with the structure and considered it a good approach centred on gaining improved insight into the UMES topics.

Most of the users testing issues have produced based on two against groups that argued and came up with two different suggestions, also various subjects made reference to suggestions without providing a rationale for such. These differences in perspectives stem from the viewpoints of subjects and their own experience, which commonly affect their view. As an example there were two groups of references up for argument: on the main interfaces or moved from the main interface as a folder on the main menu under each topic. Furthermore, there were two groups of participant against each other in the use of colours on the interface design. The group supported the use of many colours whilst the second group supported the use of simple and formal colours (black and grey). This clearly shows that the study results are influenced by the mind-set of the developers. In order to avoid this, it was considered in the dEv resource that developers need to apply the UEMS on their products at the earliest stage of the development process and gather a larger number of views from the end user. This is one way of affecting the mind-set of the developers and allowing them to utilise all usability suggestions and remain with the majority. Accordingly, the subsequent version of the resource will adopt the recommendations agreed upon by most of the user testers in mind of achieving a product that is easy to use. Accordingly, the primary garnering of user feedback data will influence the decision-making of the developer for future development planning.

Throughout the user testing stage, various recommendations were made by the participants of the study. The various arguments between the subjects in various situations emphasises the need to complete a usability evaluation. Moreover, many of the studies supported the view that software developers are able to complete the usability assessment approaches [33]. The dEv learning resource was satisfied amongst 90 % of them as usable learning resource, meaning it is recognised as a viable way of improving knowledge amongst developers concerning the usability assessment methods towards achieving a usable software. Furthermore, this study sought to evaluate the first version of the resource. This evaluation adopted a number of the evaluation strategies presented on the dEv resource, such as thinking aloud and questionnaires. The presenting results of the work emphasise how the preliminary stage of user involvement is essential and have a notable influence on the decisions of the developer in achieving usable software.

In an effort to improve the understanding of decision making and usability amongst the participants, two factors need to be taken into account, namely trust and credibility. Accordingly, credibility and trust are two important factors where learning resources should be met to allow the user to continue with the information. The dEv learning resource is concerned with these two factors in mind of encouraging users to use the product as a learning resource. Thus, the references of each topic have been placed as part of the main interface for each topic. During the user testing, participants were in complete agreement about the usefulness of including these references on the learning resource. Otherwise stated, subjects showed confidence in regard to providing content that also influences them to revisit for more resources later on.

The assistant tools and models also are helpful in increasing user knowledge on the related topics and decision making [43, 51, 52]. Moreover, some individuals sought to expand data so as to garner more information about the dEv content, with some also investing time into creating a new suggested structure for video visualisation and seeking to provide a utility navigation feature. Furthermore, when the subjects devised the approach (+,-) in mind of decreasing the content length, it became apparent that the dEv model is a way of improving the overall understanding of usability. Moreover, this emphasises that they are able to make decisions in regard to design usability.

6   Conclusion

The key element in the design of software is usability, with usability improvement also fundamental in the development process. Developers undergoing training on UEMS is essential so as to ensure their decision-making concerning software usability is encouraged. Nonetheless, creating such learning resources is not a simple task. In this work, we examined the effects of learning resources (dEv) on software usability, as well as on the usability decisions of developers. Accordingly, the resource was designed in line with particular specifications garnered throughout prior works. This study carried out various evaluation methods in an effort to gather the study data.

This study provides a key contribution concerning the research area's knowledge by creating a learning resource that encourages developers in the application of UEMS and to enhance the decision-making of usability amongst developers. The study findings emphasizes that eh learning resource influences software developers on the general usability perspective. More specifically, developers are well positioned to complete usability assessments on their products and make choices in regard to their overall usability. Learning resources that are designed in line with particular requirements may have an influence on UEMS importance and usability understanding. Furthermore, the study results emphasis the fact that the involvement of users in the earlier phases are fundamental when seeking to ensure the usability of the software is improved. The study findings are important for the interested researchers in terms of developing understanding that developers really are in need of prompting and applying UEMS by themselves, and are able to make decisions on usability. This will lead researchers to work hard in this area in order to increase developers' overall ability to create usable software. Furthermore, the study findings also are important for system designers as this can lead them to be concerned about software evaluation concepts and increase their confidently in evaluating their products so as to achieve usable design.

One of the study's main limitations is the fact that 10 participants in the UK were responsible for the empirical data. Such a small sample therefore means the findings cannot be generalised; therefore, subsequent works should make use of a larger sample of subjects in different industries so as to establish a more in-depth learning resource that is able to promote UEMS-related understanding amongst developers, in addition to their overall decision-making capacity.

References

1. Ramli, R.B.M., Jaafar, A.B.: e-RUE: A cheap possible solution for usability evaluation. In: 2008 International Symposium on Information Technology pp. 1–5. IEEE (2008)
2. Han, S.H., Yun, M.H., Kwahk, J., Hong, S.W.: Usability of consumer electronic products. Int. J. Ind. Ergon. **28**, 143–151 (2001)
3. Ferre, X., Juristo, N., Windl, H., Constantine, L.: Usability basics for software developers. IEEE Softw. **18**, 22–29 (2001)
4. Stone, D., Woodroffe, M., Minocha, S., Jarrett, C.: User Interface Design and Evaluation. Morgan Kaufmann, New York (2005)
5. Dix, A., Finlay, J., Abwod, G.D., Beale, R.: Human Computer Interaction. Prentice Hall, Upper Saddle River (2004)
6. Preece, J., Rogers, Y., Sharp, H.: Interaction Design: Beyond Human-Computer Interaction. Wiley, New York (2002)
7. Brézillon, P., Borges, M.R.S., Pino, J.A., Pomerol, J.-C.: Lessons learned from three case studies. J. Decis. Syst. **17**, 27–40 (2008)
8. Ardito, C., Costabile, M.F., Marsico, M.De, Lanzilotti, R., Levialdi, S., Roselli, T., Rossano, V.: An approach to usability evaluation of e-learning applications. Univers. Access Inf. Soc. **4**, 270–283 (2006)
9. Almansour, F., Stuart, L.: Promoting the use of design evaluation techniques within software development. In: BCS HCI (2014)
10. Nielsen, J.: Usability Engineering. Morgan Kaufmann, San Francisco (1993)
11. Shneiderman, B., Plaisant, C., Cohen, M., Jacobs, S.: Designing the user interface: pearson new international edition: strategies for effective human-computer interaction. Pearson Education Limited, New York (2013)
12. Kujala, S.: User involvement: a review of the benefits and challenges. Behav. Inf. Technol. **22**, 1–16 (2003)
13. Ardito, C., Buono, P., Caivano, D.: Investigating and promoting UX practice in industry: an experimental study. Int. J. Hum. Comput. Stud. **72**, 542–551 (2014)
14. Bowler, L., Koshman, S., Oh, J.S., He, D., Callery, B., Bowker, G., Cox, R.: Issues in user-centered design in LIS. Libr. Trends. **59**, 721–725 (2011)
15. Jokela, T., Iivari, N., Matero, J., Karukka, M.: The standard of user-centered design and the standard definition of usability: analyzing ISO 13407 against ISO. In: Proceedings of the Latin American Conference on Human-Computer Interaction, pp. 53–60. ACM (2003)
16. Rannikko, P.: User-Centered Design in Agile Software Development (2011)
17. Maguire, M.: Methods to support human-centred design. Int. J. Hum. Comput. Stud. **55**, 587–634 (2001)
18. Vredenburg, K., Smith, P.W., Carey, T., Mao, J.-Y.: A survey of user-centered design practice. In: Proceedings of the SIGCHI Conference on Human Factors in Computing Systems Changing Our World, Changing Ourselves – CHI 2002, pp. 471–478 (2002)
19. Abras, C., Maloney-Krichmar, D., Preece, J.: User-centered design. In: Bainbridge, W. (ed.) Encyclopedia Human-Computer Interact, 37th edn, pp. 445–456. Sage Publications, Thousand Oaks (2004)
20. Keinonen, T.: Protect and appreciate – notes on the justification of user-centered design. Int. J. Des. **4**, 17–27 (2010)
21. Gould, S., Powell, P.: Understanding organisational knowledge. J. Decis. Syst. **13**, 183–202 (2004)

22. Bevan, N.: Cost-benefit framework and case studies. In: Bias, R.G., Mayhew, D.J. (eds.) Cost-Justifying Usability an Update for an Internet Age, pp. 575–600. Elsevier, New York (2005)
23. Humayoun, S.R., Dubinsky, Y., Catarci, T.: A Three-Fold Integration Framework to Incorporate User–Centered Design into Agile Software Development. In: Kurosu, M. (ed.) HCD 2011. LNCS, vol. 6776, pp. 55–64. Springer, Heidelberg (2011)
24. Salah, D.: A framework for the integration of user centered design and agile software development processes. In: Proceeding of the 33rd International Conference on Software Engineering - ICSE 2011, p. 1132. ACM Press, New York, USA (2011)
25. Fox, D., Sillito, J., Maurer, F.: Agile methods and user-centered design: how these two methodologies are being successfully integrated in industry. In: Agile 2008 Conference, pp. 63–72. IEEE (2008)
26. Najafi, M., Toyoshiba, L.: Two Case Studies of User Experience Design and Agile Development. In: Agile 2008 Conference. pp. 531–536. IEEE (2008)
27. Sharp, H., Robinson, H., Segal, J.: Integrating user-centred design and software engineering: a role for extreme programming? In: BCS-HCI group's 7th Educators Workshop on Effective Teaching and Training in HC, pp. 1–4 (2004)
28. Marc, M.: 7 Benefits of agile and user centered design. https://www.thoughtworks.com/insights/blog/agile-and-user-centered-design
29. Rosenbaum, S., Rohn, J.A., Humburg, J.: A toolkit for strategic usability. In: Proceedings of the SIGCHI Conference on Human factors in Computing Systems – CHI 2000, pp. 337–344. ACM Press, New York, USA (2000)
30. Howarth, J., Smith-Jackson, T., Hartson, R.: Supporting novice usability practitioners with usability engineering tools. Int. J. Hum. Comput. Stud. **67**, 533–549 (2009)
31. Skov, M.B., Stage, J.: Supporting problem identification in usability evaluations. In: Proceedings of 17th Australia Conference on Computer-Human Interaction: Citizens Online: Considerations for Today and Computer-Human Interaction Special Interest Group, pp. 1–9 (2005)
32. Bruun, A., Stage, J.: Barefoot usability evaluations. Behav. Inf. Technol. **33**, 1148–1167 (2014)
33. Skov, M., Stage, J.: Training software developers and designers to conduct usability evaluations. Behav. Inf. Technol. **31**, 425–435 (2012)
34. Holzinger, A.: Usability engineering methods for software developers. Commun. ACM **48**, 71–74 (2005)
35. Nielsen, J., Mack, R.L.: Usability Inspection Methods. Wiley, New York (1994)
36. Gould, J.D., Lewis, C.: Designing for usability: key principles and what designers think. Commun. ACM **28**, 300–311 (1985)
37. Bargas-Avila, J.A., Lötscher, J., Orsini, S., Opwis, K.: Intranet satisfaction questionnaire: development and validationof a questionnaire to measure user satisfaction with the intranet. Comput. Human Behav. **25**, 1241–1250 (2009)
38. Nielsen, J.: Quantitative studies: how many users to test?. https://www.nngroup.com/articles/quantitative-studies-how-many-users/
39. Johnson, R.B., Onwuegbuzie, A.J.: Mixed methods research: a research paradigm whose time has come. Educ. Res. **33**, 14–26 (2004)
40. Sandelowski, M.: Focus on research methods combining qualitative and quantitative sampling, data collection, and analysis techniques. Res. Nurs. Health **23**, 246–255 (2000)
41. Hwang, W., Salvendy, G.: Number of people required for usability evaluation: the 10 ± 2 rule. Commun. ACM **53**, 130–133 (2010)
42. Nielsen, J.: Why you only need to test with 5 users. https://www.nngroup.com/articles/why-you-only-need-to-test-with-5-users/

43. Aberg, J., Shahmehri, N.: The role of human web assistants in e-commerce: an analysis and a usability study. Internet Res. **10**, 114–125 (2000)
44. Boiano, S., Bowen, J., Gaia, G.: Usability, design and content issues of mobile apps for cultural heritage promotion: The Malta culture guide experience. arXiv Preprint (2012). arXiv1207.3422
45. Burrell, A., Sodan, A.C.: Web Interface Navigation Design: Which Style of Navigation-Link Menus Do Users Prefer? In: 2006 Proceedings of 22nd International Conference on Data Engineering Workshops, pp. 1–10 (2006)
46. Magazine, S.: Breadcrumbs In Web Design: Examples And Best Practices, http://www.smashingmagazine.com/2009/03/breadcrumbs-in-web-design-examples-and-best-practices/
47. Johnson, T.: Evaluating the Usability of Collapsible Sections (or jQuery's Content Toggle), http://idratherbewriting.com/2013/03/25/evaluating-the-usability-of-collapsible-sections-or-jquerys-content-toggle/#comment-2029521352
48. Alkhuraiji, A., Liu, S., Oderanti, F.O., Megicks, P.: New structured knowledge network for strategic decision-making in IT innovative and implementable projects. J. Bus. Res. (2015)
49. Farrell, S.: Utility navigation: what it is and how to design it. http://www.nngroup.com/articles/utility-navigation/
50. Mayer, R.E., Moreno, R.: Nine ways to reduce cognitive load in multimedia learning nine ways to reduce cognitive load in multimedia learning. Educ. Psychol. **1520**, 43–52 (2010)
51. Lutz, M., Boucher, X., Roustant, O.: Methods and applications for IT capacity decisions: bringing management frameworks into practice. J. Decis. Syst. **22**, 332–355 (2013)
52. Alkhuraiji, A., Liu, S., Oderanti, F.O., Annansingh, F., Pan, J.: Knowledge network modelling to support decision-making for strategic intervention in IT project-oriented change management. J. Decis. Syst. **23**, 285–302 (2014)

How to Support Decision Making of Local Government in Prioritising Policy Menu Responding to Citizens' Views: An Exploratory Study of Text Mining Approach to Attain Cognitive Map Based on Citizen Survey Data

Hiroko Oe[1(✉)], Yasuyuki Yamaoka[2], and Eizo Hideshima[3]

[1] The Business School Bournemouth University, Bournemouth, UK
hoe@bourunemouth.ac.ul
[2] Comparison Region Research Center, The Open University, Chiba, Japan
monthill@jf7.so-net.ne.jp
[3] Nagoya Institute of Technology, Nagoya, Japan
hideshima.eizo@nitech.ac.jp

Abstract. It has been on the political agenda for the local governments how to satisfy their citizens to enhance their commitment and contribution to the communities. Especially in this ageing population era with tight fiscal conditions, it is essential for the government to know the prioritised policy menu in realising citizen satisfaction. This study aims to explore an applicable system based on citizen survey result. In our study, following literature review, we conducted focus group discussions to explore citizens' willingness to participate in local policy design, which leads us to be convinced that some activated citizens are supportive to the local governmental policy decision. Based on this qualitative result, we tried to make a cognitive map which indicated which policy fields are prioritised by citizens. Throughout this procedure, we validate the feasible practice to support local governmental decision making based on the result of citizen survey.

Keywords: Local government · Citizen perception · Citizen survey · Text mining · Cognitive map

1 Introduction

1.1 Background

Japan has been suffering from various political problems merged from depopulation, imbalanced bipolarisation, old deterioration of social infrastructure, aggravation of regional financial difficulties and other factors [1]. Especially the social infrastructure maintenance and other long-term social policy require huge amount of budget and its period of gestation is longer, hence the local governments should put priorities in policy agenda.

© Springer International Publishing Switzerland 2016
S. Liu et al. (Eds.): ICDSST 2016, LNBIP 250, pp. 202–216, 2016.
DOI: 10.1007/978-3-319-32877-5_15

Ministry of Land, Infrastructure, Transport and Tourism [2] provided some suggestions for the local governments and relevant stakeholders to combine their resources to check the policy framework so that they can maintain sustainability in designing their communities, by enhancing citizens participation and contribution, responding to this, many governments have been trying to gather citizens voices via surveys to re-design their community managerial schemes. But still they seem to have difficulties in finding the antecedent factors in determining citizen satisfaction which could lead to their participation and contribution to the communities. This means that the budget expended to conduct citizen survey is not utilised which has been on the policy agenda for a while [3, 4].

The discussion which policies and public activities by the local governments are appreciated by the citizens and leads to the citizen satisfaction is missing [4–8]. Indeed quite a few local councils are interested in conducting citizen survey, however the best practice how to reflect its result in the real policy making phase are still under experiment [4, 5, 9–11]. This is the background for this study. Hence, it has been arranged to support governmental decision making through data mining, which indicates more effort in realising its favourable and effective outcome [12–18].

1.2   Aim and Objectives of This Study

The aim of this study is to find the feasible procedure for the local governmental decision making in prioritising policy menu based on citizen survey data. Hence, we, in the first place, investigate the citizens' willingness to support their local government as active participants by conducting focus group discussion (FGD) to depict essential views from citizen's intention to participate and contribute to their local governmental policy making. Based on the essence from FGDs, we build a cognitive map from the data of citizen survey to indicate potential pathway in prioritising policy menu. A cognitive map is built using a commercial text mining system which is affordable for the public sector rather than implementing a gigantic data analytical software. We expect from this reasonably designed two steps of data analysis to propose a realistic action plan for the local government to utilise the result of citizen survey, which means that the budget for the survey itself could be effectively valued which is another aspect for the good practice of local government [20, 21].

Hence, our objectives are:

(a) to explore the citizen's perceptions towards local government activities,
(b) to attain a cognitive map based on commercial text mining system,
(c) to evaluate the effectiveness of the b) as a data mining to support local governments' decision making,
(d) to present good practical implication for the local governments who have been seeking the best way in decision making of prioritised policy menu.

2 Literature Review

2.1 Overall Discussion

Crowley and Coffey [22] pointed out that Bridgman and Davis [23] have argued that 'ideally government will have a well developed and widely distributed policy framework, setting out economic, social and environmental objectives', hence local governments should aim at reconnecting with community priorities and at redirecting macropolicy setting away from a preoccupation with economic priorities, respectively. Irimia [24] analyses the political framework and citizens satisfaction based on a socio-cultural perspective and found we can come closer to the intangible aspects and hidden antecedent factors realizing their satisfaction. Then next question should be how to prioritise policy menu among various dimensions of those. Also how to support the best decision making to satisfy citizens in maintain them loyal to the government has been on the agenda up today [12, 19].

2.2 Satisfied Citizens and Positive Support Towards the Local Governments

There are also discussions about citizens' psychological aspects also for the debate of their satisfaction. Cordes et al. [25] investigate the application of contingent valuation in social capital, attachment value and rural development. Crawford et al. [26] debate the need for the value of community participation [5, 25–31]. Robison and Flora [33] present a social capital paradigm that includes social capital, socio-emotional goods, attachment values, networks, institutions and powers. Julian and Ross [34] explore infrastructure and functions to serve as a foundation for collaborative community problem solving. Yoshitake and Deguchi [35] analyse a Japanese case of the town planning in the depopulation era to indicate that the residents' community attachment and human encounters, mutual understanding and discreet leadership could be essential. They evaluate this aspect of citizens' perception as a positive inclination and supportiveness towards their own local government [17, 36].

Regarding citizen participation and their satisfaction, Bedford et al. [37] debate the extent to which collaborative approaches to planning and decision making are capable of giving precedence to the concerns of the public or of promoting trust in local institutions, indicating that the possibility of new participation practices associated with the development control process to determine citizens satisfaction level. Noworól [38] contribute to an understanding of the influence of social capital on the contemporary community management, also Olsson [39] discuss why and how to lead local citizens and their values into account in urban management and planning. Similarly, Martin and Morehead [40] discuss the importance of civic governance and civic infrastructure, implying that civic engagement, inclusiveness, leadership, participation, community vision, diversity and networking as elements are fundamentals for the civic governance in realising citizens satisfaction. Perhaps we could name this 'civic infrastructure'.

2.3   Health and Welfare Policies and Satisfaction in the Ageing Era

Warburton et al. [41] debate that effective policy responses to the ageing of the population are a priority area for government and non-government agencies, Williams and Pocock [42] also debate how to cope with population portfolio changes which had a major impact especially social policies such as health and welfare matters. It is inevitable for us to discuss health and welfare policies and citizen satisfaction in this ageing era [43–45]. Caldwell et al. [44] indicate the concept of the ageing population is one that has generated much debate and discussion at global, national and local levels. Julian and Ross [34] also debate researchers should focus on problem solving in the health, education, and social services arenas. Mitchell and Shortell [46] introduce the concept 'community health partnerships' as voluntary collaborations of diverse community organisations in order to pursue a shared interest in improving community health. Here again we can see the area which citizens' views should be reflected to collaborate in choosing a good policy framework [7, 47, 48].

2.4   Social Infrastructure Improvement and Its Decision Making

Social infrastructure issues should be another factor influencing citizens' satisfactory level. Work, Patrick and Roseland [49] analyse the importance of developing community in the residential areas, indicating that specific physical and social infrastructure in the residential area and the workplace should be prioritised on the political agenda. Dabinett [50] analyse urban areas where we should pursue a range of regeneration policies. Marshall [51] shows how distinctive the UK context has become since the 1990s, with the new legislation emerging from the political economy of powerful infrastructure industries, interacting with the particular configuration of interests represented by the New Labour government. Similarly also in Australia, as McShane [36] debated the importance of a management of community infrastructure by local governments, coping with a shortfall to maintain ageing physical assets. However, the policy seems to rely on a model of local government as a service provider only that is inattentive to new interests in community building and governance and takes limited account of the wider social value of community facilities [29, 36, 45, 52]. This means that the issue how and to what extent local governments should implement its infrastructure refurbishment is on the policy agenda. Decision support procedure is required [40, 53].

2.5   Environmental Beautification

The improvement of rural or city environment have been oriented toward raising the level of living standard for the citizens, since 1970's in Japan especially in Japan for local towns and cities after the high economic growth [54]. The economic growth caused some issues such as the surplus of farm products, the air and water pollutions, heavy traffics on roads, and nature destructions in rural areas [54, 55]. The movement for beautification of city environment has risen since 1980, affected by the German movement to 'make our village beautiful' [54]. Then next question should be the factor of 'beautification' could enhance citizen's attachment or loyalty to their local area or not

[56, 57]. Fransico [57] proves that willing to pay (WTP) for an improvement in urban aesthetic values are significant, even he indicates that WTP for urban aesthetic improvements increases more than proportionally as income rises, which implies that improving city aesthetics are valued by citizens. This debate is compatible with Shigeoka's [54] argument that community aesthetics has developed to bring up the love for their own homeland.

2.6 City Buzz

Age-friendly city components are presented by Kennedy [58], whereas he emphasises the importance of re-imagining city with aging-in-place communities should be vibrant enough to attract young people, while welcoming older people to stay in their same community for a lifetime. Ghahramanpouri et al. identify contributing factors for urban social sustainability, indicating that town vibrancy could be the basis in the aging era. Similarly scholars debate the impact of city buzz and vibrancy as one of the factors to attract people to vitalise the relevant communities which could contribute to the sustainability of the area [60–62].

There are several research and case studies to explore factors which could enhance the city attractiveness. They implied the importance of city buzz as an origin for the sustainability with loyal citizens to the area [62–68]. In this context some scholars indicated the role of social entrepreneurs in the area to generate community associations and networking that could produces favourable social outcomes [69].

2.7 Trust for Local Government

Trust from citizens to the local government is critical for managerial flexibility and political accountability in the modern administrative state [30]. There is an accumulation of research focusing on the impact of trust which leads to good support for both central and local governmental activities and to enhance further citizen commitment and their public involvement [70–73]. Moreover quite a few scholars have explored the impact of trust as an antecedent factor to influence citizen satisfaction towards their governments [11, 74, 75].

2.8 Research Questions

Following the literature review above, the research is designed to find out how citizens evaluate and perceive governmental policy making in their local government and to provide explanations of why they regard particular issues important in the context of their citizenship. To answer this question, we explore the citizens' willingness to participate and support their local governmental policy designing, which is going to be followed by the evaluation of the cognitive map attained from the citizen survey result.

3 Methodology

This research is conducted based on two phases, both are qualitative approaches; FGD and cognitive mapping.

The academic interest in the decision-making processes is concerned with human behaviours. In this context, to understand what is happening, it is important to focus on meanings [76]. Focus groups were chosen as being an appropriate methods of eliciting meanings [77]. The method is also efficient, being capable of obtaining a good set of citizens' views in an efficient way. Inductive approaches were taken so that the views attained from FGD were those of the citizens who are willing to present their opinions in a supportive way, which are not determined by the prescriptive approaches that are conducted in positivist research [77].

Cognitive mapping is a qualitative technique designed to identify cause and effect as well as to explain causal links [77]. It is used to structure, analyse and make sense of problems [77]. There are a small number of computer programs that can be used to conduct this type of work. As our aim is to evaluate the applicability of supportive decision making software, so as a material for doing it, we conduct text mining based on a relevant commercialised software to attain a cognitive map of informants.

3.1 FGD and Data Gathering

18 citizens aged 35 and over volunteered to participate in our FGDs, they were divided into three groups and facilitated by one of the authors for active collaborative discussions. [77] presents the importance that researcher's notion of what a focus group is will influence the quality and therefore the utility of responses received, categorising in three types of focus groups which are phenomenological interaction-centred focus groups, clinical focus groups and exploratory research. An exploratory research being most common to examine a topic quickly and at low cost before conducting follow-up survey [77].

Participants to our FGD are diverse with regard to demographic characteristics, and they are asked to respond to three general questions:

- Are you willing to contribute to and support the policy making of your local government?
- What do you perceive the barriers to implement effective public participation?
- How can you evaluate the current citizen participation in prioritising the best portfolio of policy menu?

Followed by free conversation and discussion among the participants facilitated by the authors with a guidance from a perspective how to support decision making with active citizens' supportive participation.

The discussions lasted for about two hours each group and were recorded and transcribed. In the first stage of the analysis, we coded the transcribed interviews and discussions individually, using a qualitative form of content analysis which was suggested by Corbin and Strauss [78]. In the second stage of analysis, we will put the data into the text mining system to obtain a cognitive mapping to create visualised categorical themes which should indicate citizens' perceptions for the prioritised policy menu.

3.2 Cognitive Mapping of the Citizens' Perception

During the FGDs, the interviews were facilitated and guided by the researcher and invited to discuss freely issues relating to the local governmental activities. As our aim is to explore the possibility to support local governmental decision making, we will create the map to be able to focus on the most prioritised areas to decide the holistic policy menu which could make their citizens satisfied with their local governments. Cognitive map has been utilised by researchers as one of the idealistic method to visualise the perceptions of interviewees to explore the best option in responding to the research questions [14, 77].

4 Analysis and Findings

4.1 FGDs Exploring Citizens' Willingness to Participate in Policy Designing

From three FGDs with a variety of participants, there are quite divers comments were collected.

Disappointment for the lack of the opportunities to participate in governmental decision making. The first category of the comments could be summarised that citizens are ready to contribute to their local governments but still they are not satisfied because of lacking of the opportunities to do so.

A: 'We would like to see more public policies are reflecting our views, however we are not feeling that our views are used by local governments which is a shame' (male, 50s).
B: 'The local government seems to have tried a lot to attract citizens to have more interest in their activities but still we are not sure how to contribute to our own government' (male, 40s).
C: 'It is nice for us to be contributors to the local governments, but being a citizen committee member is too difficult for our age as we have so many other thing to do such as growing our small kids' (female, 30s).

The utilisation of the result of citizen survey. The second category could be summarised as the result of citizen survey.

D: 'The local government has conducted several citizen surveys, what happened to the result, we are not sure....' (male, 40s).
E: 'It is common for the public sectors to conduct some kinds of surveys, but the conduct itself is now their aim, what happened to the result of those? We seldom hear about it. Strange!' (male, 30s).
F: 'How could they make the most of the data of the survey? The bureaucratic organisations usually fail to make the most of it!' (male, 60s).

The self-awareness as tax payers

G: 'Of course it is their duty and responsibility to make it accountable how and what they decide based on our survey data, as everything was paid by our tax' (male, 50s). H: 'As a tax payer, it is our responsibility to contribute in decision making of the local government, but it is not clear by which occasion and procedure we can do so....' (female, 30s).

The barriers preventing citizens' support in decision making of local government. Final category could be summarised as 'their perception of the barriers for the citizens to support governmental decision.

I: 'Silent majority. Not all of us would like to contribute to our own local government. The majority of the citizens are just sitting there, as free riders' (male, 50s) J: 'The structure of the organization. It seems to be difficult for the government to share the views and opinions from us' (female, 60s). K: 'The lack of communication between us. Some of the citizens are active and willing to support the government, however, the good conduit between us and the government is not robust enough to encourage our interaction towards the good collaboration to realise good policy menu' (male, 40s).

The outcome of FGDs. As overviewed in the previous sections, citizens seem to have a doubt that their precious views through citizen survey is not utilised enough, although they are sincerely willing to contribute in supporting the governmental decision. Moreover, the result of the survey is not visualised clearly and feed backed to the citizens, which might affect their motivation in being active to support their government.

The outcome of the FGDs indicate that it is essential for the local government to feed back the result of the relevant survey to the citizens and in doing so, citizens' motivation to contribute to the local governmental activities could be enhanced and through this path the trust between citizens and governments could be nurtured which could build the good basis as a platform on which all of the stakeholders of the government could collaborate in designing ideal policy menu.

4.2 Building a Cognitive Map Based on Text Mining

Following the FGDs, we now are going to the next stage to build a cognitive map using the result of citizen survey. To make the result visualised could be effective for enhancing the understanding level of the survey result among citizens, at the same time, it could be useful to evaluate the holistic views which policy activities are prioritised by citizens which could be a good material for the governmental decision making.

Procedure. To attain a cognitive map, we implement SPSS text analytics for survey version 4.0.1. There are several software available, but as an experimental basis, we decided to use this as we usually analyse the data quantitatively based on SPSS systems.

Figure 1 shows the process of data preparation on this system. We use 1204 comments from survey participants.

Fig. 1. Text mining procedure

Setting categories. Before starting text mining, we are required to set categories to analyse the dataset. Figure 2 shows the screenshot of the setting. In this case, we analyse all of the 'nouns', then categorise 'good/bad', 'satisfied/unsatisfied', 'bad/worrying/anxiety', 'please/ask', 'proposition/warn' and 'trouble/improvement'. SPSS text analytical protocol is eligible for the unique Japanese linguistic characteristics, hence we also check our analysis will consider Japanese language context as well.

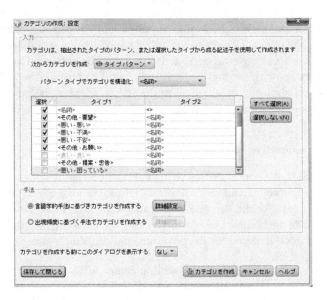

Fig. 2. Categories setting

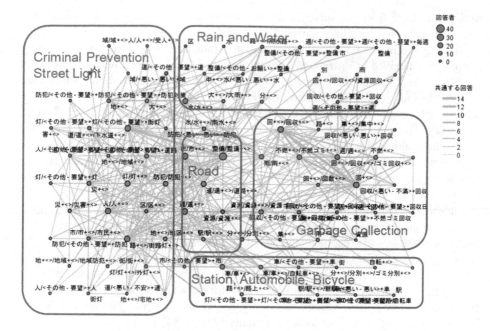

Fig. 3. Grid and categorization

4.3 Grid and Categorisation

Following this preparation, we attain grid layout of the key words and Fig. 3 shows our categorisations with red circles. Grid layout indicates which words are close and which are frequently appeared in the comments of the survey participants.

Based on the information of this grid layout, we categorise five groups as the policy menu which citizens perceive important to implement; 'security on the street at night', 'water and flooding prevention', 'road construction', 'collect the rubbish' and 'parking lots in the station'.

5 Discussion and Conclusion

5.1 Discussion

From our analyses of the first phase as FGDs of activated participants, there are several key words were depicted. Among those, we could estimate those who are willing to contribute to the local government have noticed some barriers in doing so because of the bureaucratic oraganisation, and also they perceive the lack of instruments by which their views are reflected in policy selection by their local government. Moreover they appreciate governmental attitudes to implement citizens' views via conducting citizen survey, however, they seem to be frustrated with the vague outcome what and how those views were utilized in the governmental actions. Hence our first phase of research can conclude that it is required for the local government to introduce some solid procedure to depict essence from the survey result and enhance communication with conductor and participants for the citizen survey.

Bearing this in mind, we built a cognitive map indicating which categories of policy actions are perceived by citizens as prioritized fields of policy menu which is shown as the Fig. 3. As it shows, there are five areas which citizens prioritised, all of those are quite daily-life based issues, quite different scenery was attained compared to the outcome of previous literature discussion which is an interesting finding. In our Sect. 2, we surveyed the relevant literature review to present some essential fields such as indicate some health and welfare policies and city beautification, however we found out other elements such as security on the streets at night, improvement of roads and parking lots of stations and prevention of flooding. In other words, citizens seem to perceive more importance on the public activities in the basic security related fields.

Indeed, social capital implication which prevailed in Cabinet Office Initiatives, Japan [79], nurturing community bonding and commitment has been discussing quite a long time since then as social capital could be helpful for building sustainable communities, however, perhaps this kind of 'social capital' debate is applicable only when basic infrastructure is solid to enable citizens feel secured in the community. Moreover, even elderly people would prefer to prioritise social infrastructure improvement rather than health and welfare policy, which could imply that we should once again re-consider the layout of policy agenda in the local community [25, 26, 34]. To be able to present this ironical gap with theoretical discussion and reality of citizens' perception could be discovered by building cognitive map based on the citizen survey result, which indicates its usefulness and applicability in supporting local governmental decision making.

5.2   Limitations and Recommendations

The content of this study could contribute in providing the feasible supporting system for the local governmental decision making. Especially considering the tight budget circumstances, it is essential for them to implement effective system to conduct reasonable policy selection. However, this research should be deepened with a specific scope how the local governments plan and conduct effective citizen survey in the first place. Otherwise, we can obtain the robust origin for our cognitive mapping to depict their perceptions useful for them to analyse and explore their real views in the community governance context. In addition, how to analyse the relations among policy menu and achievable political goals 'how to realise citizen satisfaction' is still on our research agenda.

In reality, when implementing the policy menu, the budget limitations and other political issues exist always there, which should be discussed at the same time. In this study, only several policy fields are discussed with relevant literature, however many other issues are also interrelated with the real policy actions holistically, hence it should be our social researchers' responsibility to support local governmental practices with provision of feasible decision support systems with trial and error experiments based on the cooperation and collaboration with stakeholders which could contribute to the local communities in the end. Through these actions, the real sustainable relationships among local government and citizens could be achieved via enhancing citizen satisfaction and their commitment, by finding antecedent factors realising their satisfaction and trust towards their local government [5, 20, 21]. The authors would like to continue the relevant study so that we could produce more robust proposal in supporting the governmental decision making in choosing the best combination of prioritised policy menu.

As another limitation, in this project, we could not enhance our research scope to policy makers perceptions of whose opinions and views are critical in integrating in the holistic proposal of this data analysis. We acknowledge this limitation as essential to be mitigated as they are responsible for policy making in most representative democracies.

Moreover, the informants of this dataset are only over 35+ years of age, which might have caused an imbalance of the data variance. Hence when we build a proposal of this procedure to the practices in the reality, we should conduct another data collection and test of the usefulness of this software based on a broader and better balanced age bands of respondents.

References

1. National Land Council, Japan (2011) "Long-term View of the Country" Interim Daft Outline http://www.mlit.go.jp/common/000135837.pdf. Accessed 26 Oct. 2015
2. Ministry of Land, Infrastructure, Transport and Tourism (2013) White Paper on Land, Infrastructure and Transport in Japan 2012. http://www.mlit.go.jp/english/white-paper/2011.pdf. Accessed 16 Oct. 2015
3. Watson, D.J., Juster, R.J., Johnson, G.W.: Institutionalized use of citizen surveys in the budgetary and policy-making processes: a small city case study. Public Adm. Rev. **51**, 232–239 (1991)
4. Herian, M.N., Tomkins, A.J.: Citizen satisfaction survey data: a mode comparison of the derived importance-performance approach. Am. Rev. Public Adm. **42**(1), 66–86 (2012)
5. Kelly, J.M., Swindell, D.: A multiple-indicator approach to municipal service evaluation: correlating performance measurement and citizen satisfaction across jurisdictions. Public Adm. Rev. **62**(5), 610–621 (2002)
6. Brown, T.: Coercion versus choice: citizen evaluations of public service quality across methods of consumption. Public Adm. Rev. **67**(3), 559–572 (2007)
7. Dowding, K., John, P.: Voice and choice in health care in England: understanding citizen responses to dissatisfaction. Public Adm. **89**(4), 1403–1418 (2011)
8. Greasley, S., John, P.: Does stronger political leadership have a performance payoff? Citizen satisfaction in the reform of subcentral governments in England. J. Public Adm. Res. Theor. **21**(2), 239–256 (2011)
9. Kelly, J.M.: The dilemma of the unsatisfied customer in a market model of public administration. Public Adm. Rev. **65**(1), 76–84 (2005)
10. Poister, T.H., Henry, G.T.: Citizen ratings of public and private service quality: a comparative perspective. Public Adm. Rev. **54**(2), 155 (1994)
11. Morgeson, F.V.: Expectations, disconfirmation, and citizen satisfaction with the US federal government: testing and expanding the model. J. Public Adm. Res. Theor. **23**(2), 289–305 (2013)
12. Lourenco, R.P., Costa, J.P.: Incorporating citizens' views in local policy decision making processes. Decis. Support Syst. **43**(4), 1499–1511 (2007)
13. Ranerup, A.: Decision support systems for public policy implementation: the case of pension reform. Soc. Sci. Comput. Rev. **26**(4), 428–445 (2008)
14. Moreno-Jiménez, J.M., Cardeñosa, J., Gallardo, C.: Arguments that support decisions in e-Cognocracy: a qualitative approach based on text mining techniques. In: Lytras, M.D., et al. (eds.) Visioning and Engineering the Knowledge Society. A Web Science Perspective. LNCS, vol. 5736, pp. 427–436. Springer, Heidelberg (2009)

15. Rao, G.K., Dey, S.: Decision support for e-Governance: a text mining approach (2011a)
16. Rao, G.K., Dey, S.: Text mining based decision support system (TMbDSS) for E-governance: a roadmap for India. In: Wyld, D.C., Wozniak, M., Chaki, N., Meghanathan, N., Nagamalai, D. (eds.) ACITY 2011. CCIS, vol. 198, pp. 270–281. Springer, Heidelberg (2011)
17. Mostafa, M.M., El-Masry, A.A.: Citizens as consumers: profiling e-government services' users in Egypt via data mining techniques. Int. J. Inf. Manage. 33(4), 627–641 (2013)
18. Fleming, L.E., Haines, A., Golding, B., Kessel, A., Cichowska, A., Sabel, C.E., Depledge, M.H., Sarran, C., Osborne, N.J., Whitmore, C., Cocksedge, N., Bloomfield, D.: Data mashups: potential contribution to decision support on climate change and health. Int. J. Environ. Res. Public Health 11(2), 1725–1746 (2014)
19. De Cnudde, S., Martens, D.: Loyal to your city? A data mining analysis of a public service loyalty program. Decis. Support Syst. 73, 74–84 (2015)
20. Stipak, B.: Citizen Satisfaction with urban services: potential misuse as a performance indicator. Public Adm. Rev. 39(1), 46–52 (1979)
21. Van Ryzin, G.G.: Testing the expectancy disconfirmation model of citizen satisfaction with local government. J. Public Adm. Res. Theor. 16(4), 599–611 (2006)
22. Crowley, K., Coffey, B.: New governance, green planning and sustainability: Tasmania together and growing victoria together. Aust. J. Public Adm. 66(1), 23–37 (2007)
23. Bridgman, P., Davis, G.: The Australian Policy Handbook. Allen and Unwin, Sydney (2000)
24. Irimia, H.: Regional Economy and Public Administration. Annals of Eftimie Murgu University Resita, Fascicle II, Economic Studies: 199–202 (2011)
25. Cordes, S., Allen, J., Bishop, R.C., Lynne, G.D., Robison, L.J., Ryan, V.D., Shaffer, R.: Social capital, attachment value, and rural development: a conceptual framework and application of contingent valuation. Am. J. Agric. Econ. 85(5), 1201–1207 (2003)
26. Crawford, P., Kotval, Z., Rauhe, W., Kotval, Z.: Social capital development in participatory community planning and design. TPR. Town Plan. Rev. 79(5), 533–553 (2008)
27. Smith, G.R.: Achieving sustainability: exploring links between sustainability indicators and public involvement for rural communities. Landscape J. 19(1/2), 179 (2000)
28. Yang, K., Callahan, K.: Citizen involvement efforts and bureaucratic responsiveness: participatory values, stakeholder pressures, and administrative practicality. Public Adm. Rev. 67(2), 249–264 (2007)
29. Arvind, G.R.: The state, local community and participatory governance practices: prospects of change. Proc. World Acad. Sci.: Eng. Technol. 46, 697–708 (2008)
30. Cooper, C.A., Knotts, H.G., Brennan, K.M.: The importance of trust in government for public administration: the case of zoning. Public Adm. Rev. 68(3), 459–468 (2008)
31. McIntyre, T., Halsall, J.: Community governance and local decision making: paper presented to the 'diversity and convergence: planning in a world of change' conference. Local Econ. 26(4), 269 (2011). (Sage Publications, Ltd.)
32. Creighton, J.L.: The Public Participation Handbook: Making Better Decisions Through Citizen Involvement, 1st edn. Jossey-Bass, San Francisco (2005). James L. Creighton [Bibliographies Handbooks Non-fiction], c2005
33. Robison, L.J., Flora, J.L.: The social capital paradigm: bridging across disciplines (Paul Wilson, University of Arizona, presiding) the social capital paradigm: bridging across disciplines. Am. J. Agric. Econ. 85(5), 1187–1193 (2003)
34. Julian, D.A., Ross, M.: Strengthening infrastructure and implementing functions to support collaborative community problem solving. J Planning Lit. 28(2), 124–134 (2013)
35. Yoshitake, T., Deguchi, C.: Social capital development in a rural community based on exchange management with outsiders. TPR: Town Plan. Rev. 79(4), 427–462 (2008)

36. McShane, I.: Social value and the management of community infrastructure. Aust. J. Public Adm. **65**(4), 82–96 (2006)
37. Bedford, T., Clark, J., Harrison, C.: Limits to new public participation practices in local land use planning. TPR: Town Plan. Rev. **73**(3), 311 (2002)
38. Noworól, A.: Social capital and multilevel territorial management. the case of Poland. Kapitał społeczny a wielopasmowe zarządzanie terytorialne. Przypadek Polski **2**, 99–110 (2011)
39. Olsson, K.: Citizen input in urban heritage management and planning. TPR: Town Plan. Rev. **79**(4), 371–394 (2008)
40. Martin, S.A., Morehead, E.: Regional indicators as civic governance: using measurement to identify and act upon community priorities. Natl. Civic Rev. **102**(1), 33–42 (2013)
41. Warburton, J., Everingham, J.-A., Cuthill, M., Bartlett, H.: Achieving effective collaborations to help communities age well. Aust. J. Public Adm. **67**(4), 470–482 (2008)
42. Williams, P., Pocock, B.: Building 'community' for different stages of life: physical and social infrastructure in master planned communities. Community Work Fam. **13**(1), 71–87 (2010)
43. Barnes, M.: Users as citizens: collective action and the local governance of welfare. Soc. Policy Adm. **33**(1), 73–90 (1999)
44. Caldwell, K., Saib, M., Coleman, K.: The ageing population: challenges for policy and practice. Divers. Health Soc. Care **5**(1), 11–18 (2008)
45. Corbett, S.: Review of the big society debate: a new agenda for social welfare? Crit. Soc. Policy **33**(1), 185–187 (2013)
46. Mitchell, S.M., Shortell, S.M.: The Governance and Management of Effective Community Health Partnerships: A Typology for Research, Policy, and Practice. The Milbank Quarterly **78**(2), 241 (2000)
47. Driedger, S.M.: Creating shared realities through communication: exploring the agenda-building role of the media and its sources in the E. coli contamination of a Canadian public drinking water supply. J. Risk Res. **11**(1/2), 23–40 (2008)
48. Jorna, F.: Going Dutch. Informatisation and citizen-government relations. Going Dutch. Informatisierung und Bürger-government relations **2014**(1), 21–30 (2014)
49. Patrick, R., Roseland, M.: Developing sustainability indicators to improve community access to public transit in rural residential areas. J. Rural Community Dev. **1**(1), 1–17 (2005)
50. Gordon, D.: Realising regeneration benefits from urban infrastructure investment: lessons from Sheffield in the 1990s. TPR: Town Plan. Rev. **69**(2), 171 (1998)
51. Marshall, T.: Reforming the process for infrastructure planning in the UK/England 1990–2010. TPR: Town Plan. Rev. **82**(4), 441–467 (2011)
52. Jakaitis, J., Stauskis, G.: Facilitating sustainable development of uyrban landscape by involvement of local territorial communities in vilnius city. Archit. Urban Plan. **5**, 105–111 (2011)
53. Lemert, C.: Review of 'Genealogies of citizenship: markets, statelessness, and the right to have rights'. Am. J. Sociol. **117**(3), 989–991 (2011)
54. Shigeoka, T.: The improvement of rural life environment and the reconstruction toward beautiful rural community in Japan. J. Rural Stud. **6**(2), 36–47 (2000)
55. Bookman, S., Woolford, A.: Policing (by) the urban brand: defining order in Winnipeg's exchange district. Soc. Cult. Geogr. **14**(3), 300–317 (2013)
56. Taylor, D.E.: Minority environmental activism in Britain: from Brixton to the lake district. Qual. Sociol. **16**(3), 263 (1993)
57. Fransico, J.P.S.: Are the rich willing to pay for beautiful cities?: income effects on the willingness to pay for aesthetic improvements. Australas. J. Reg. Stud. **16**(2), 233 (2010)
58. Kennedy, C.: The City of 2050 - an age-friendly, vibrant, intergenerational community. Generations **34**(3), 70–75, 76 (2010)

59. Ghahramanpouri, A., Abdullah, A.S., Sedaghatnia, S., Lamit, H.: Urban social sustainability contributing factors in Kuala Lumpur streets. Procedia – Soc. Behav. Sci. **201**, 368–376 (2015)
60. Daskon, C., McGregor, A.: Cultural capital and sustainable livelihoods in Sri Lanka's rural villages: towards culturally aware development. J. Dev. Stud. **48**(4), 549–563 (2012)
61. Harun, N.Z., Zakariya, K., Mansor, M., Zakaria, K.: Determining attributes of urban plaza for social sustainability. Procedia – Soc. Behav. Sci. **153**, 606–615 (2014)
62. Nzeadibe, T.C., Mbah, P.O.: Beyond urban vulnerability: interrogating the social sustainability of a livelihood in the informal economy of Nigerian cities. Rev. Afr. Polit. Econ. **42**(144), 279–298 (2015)
63. Glazer, N.: Shanghai surprise- the most vibrant city in the world isn't New York. New Repub. **238**(1), 12–13 (2008)
64. Baker, P.: LONDON LORE the legends and traditions of the world's most vibrant city. TLS-Times Lit. Suppl. **5518**, 28 (2009)
65. Rizzi, P., Dioli, I.: Strategic planning, place marketing and city branding: the Italian case. J. Town City Manag. **1**(3), 300–317 (2010)
66. Macleod, G., Johnstone, C.: Stretching urban renaissance: privatizing space, civilizing place, summoning 'community'. Int. J. Urban Reg. Res. **36**(1), 1–28 (2012)
67. Wetzstein, S.: Globalising economic governance, political projects, and spatial imaginaries: insights from four Australasian1 cities globalising economic governance, political projects, and spatial imaginaries: insights from four Australasian cities. Geogr. Res. **51**(1), 71–84 (2013)
68. Anderson, T.: City profile: malmo: a city in transition. Cities **39**, 10–20 (2014)
69. Ardichvili, A., Cardozo, R., Ray, S.: A theory of entrepreneurial opportunity identification and development. J. Bus. Ventur. **18**, 105–123 (2003)
70. Garcia, G.A., Gaytan, E.A.: Trust in Spanish governments: antecedents and consequences. Econ. Anal. Policy **43**(2), 177 (2013)
71. Myung, J.: Citizen participation, trust, and literacy on government legitimacy: the case of environmental governance. J. Soc. Change **5**(1), 11–25 (2013)
72. Smith, J.W., Leahy, J.E., Anderson, D.H., Davenport, M.A.: Community/agency trust and public involvement in resource planning. Soc. Nat. Resour. **26**(4), 452–471 (2013)
73. Herian, M.N., Shank, N.C., Abdel-Monem, T.L.: Trust in government and support for governmental regulation: the case of electronic health records. Health Expect. **17**(6), 784–794 (2014)
74. Van Ryzin, G.G., Muzzio, D., Immerwahr, S., Gulick, L., Martinez, E.: Drivers and consequences of citizen satisfaction: an application of the American customer satisfaction index model to New York City. Public Adm. Rev. **64**(3), 331–341 (2004)
75. Kathi, P.C., Cooper, T.L.: Connecting neighborhood councils and city agencies: trust building through the learning and design forum process. J. Public Aff. Educ. **13**(3/4), 617–630 (2007)
76. Geren, K.J.: New governance, green planning & sustainability: reviewing Tasmania together and growing victoria together. Aust. J. Public Adm. **66**(1), 23–37 (1988)
77. Hines, T.: An evaluation of two qualitative methods (focus group interviews and cognitive maps) for conducting research into entrepreneurial decision making. Qual. Mark. Res.: Int. J. **3**(1), 7–16 (2000)
78. Corbin, J., Strauss, A.: Grounded theory research: procedures, canons, and evaluative criteria. Qual. Sociol. **13**(1), 3–21 (1990)
79. Cabinet Office, Japan: Social Capital: Seeking a virtuous cycle of the rich human relations and civil activities (2003). http://www5.cao.go.jp/seikatsu/whitepaper/h19/01_honpen/index.html. Accessed 16 Oct. 2015

Author Index

.

Printed in the United States
By Bookmasters